OXFORD STUDIES
IN
NUCLEAR PHYSICS

GENERAL EDITOR
P. E. HODGSON

Quarks Bound by Chiral Fields

The quark structure of the vacuum
and of light mesons and baryons

GEORGES RIPKA

Senior Research Physicist,
Centre d'Etudes de Saclay,
France

CLARENDON PRESS • OXFORD

1997

Oxford University Press, Great Clarendon Street, Oxford OX2 6DP
Oxford New York
Athens Auckland Bangkok Bogota Bombay Buenos Aires
Calcutta Cape Town Dar es Salaam Delhi Florence Hong Kong
Istanbul Karachi Kuala Lumpur Madras Madrid Melbourne
Mexico City Nairobi Paris Singapore Taipei Tokyo Toronto
and associated companies in
Berlin Ibadan

Oxford is a trade mark of Oxford University Press

Published in the United States
by Oxford University Press Inc., New York

A catalogue record for this book is available from the British Library

Library of Congress Cataloging in Publication Data
(Data available)
ISBN 0 19 851784 X

Typeset by the author
Printed in Great Britain by
Biddles Ltd., Guildford, Surrey

OXFORD STUDIES IN NUCLEAR PHYSICS

General editor: P. E. Hodgson

1 J. L. McL. Emmerson: *Symmetry principles in particle physics* (1972)

2 J. M. Irvine: *Heavy nuclei, superheavy nuclei, and neutron stars* (1975)

3 I. S. Towner: *A shell-model description of light nuclei* (1977)

4 P. E. Hodgson: *Nuclear heavy ion reactions* (1978)

5 R. D. Lawson: *Theory of the nuclear shell model* (1980)

6 W. E. Frahn: *Diffractive purposes in nuclear physics* (1985)

7 S. S. M. Wong: *Nuclear statistical spectroscopy* (1986)

8 N. Ullah: *Matrix ensembles in the many-nucleon problem* (1987)

9 A. N. Antonov, P. E. Hodgson, and I. Zh. Petkov: *Nucleon momentum and density distributions in nucleii* (1988)

10 D. Bonatsos: *Interacting boson models of nuclear structure* (1988)

11 H. Ejiri and M. J. A. de Voigt: *Gamma-ray and electron spectroscopy in nuclear physics* (1989)

12 B. Castel and I. S. Towner: *Modern theories of nuclear physics moments* (1990)

13 R. F. Casten: *Nuclear structure* (1990)

14 I. Zh. Petkov and M. V. Stoitsov: *Nuclear density function theory* (1991)

15 I. Gadioli and P. E. Hodgson: *Pre-equilibrium nuclear reactions* (1992)

16 J. D. Walecka: *Theoretical nuclear and subnuclear physics* (1995)

17 D. N. Poenaru and W. Greiner (eds.): *Handbook of nuclear properties* (1995)

18 P. Fröbrich and R. Lipperheide: *Theory of nuclear reactions* (1996)

19 Fl. Stancu: *Group theory in subnuclear physics* (1996)

20 S. Boffi, C. Giusti, F. D. Pacati, and M. Radici: *Electromagnetic response of atomic nuclei* (1996)

21 G. Ripka: *Quarks bound by chiral fields* (1997)

Strange things may be generally accounted for if their cause be fairly searched out.

Jane Austen

PREFACE

This book describes dynamical models which account for the spontaneous chiral symmetry breaking in the vacuum, for the mesons which are normal vibrational modes of the vacuum and for baryons which appear as bound states of quarks (solitons). At the present time, we are still unable to construct models which account for the dynamics of both confinement and the spontaneous chiral symmetry breaking in the vacuum. Quark models which use confining potentials make use of constituent quark masses which are input parameters. Chiral models which treat spontaneous chiral symmetry breaking dynamically do account for the constituent quark masses, but they are unable to confine the quarks. Failing this, they can at least bind quarks so as to form hadrons. In this book, we describe how quarks are bound by chiral fields and to what extent such bound states can describe the observed hadrons.

Attempts to describe hadrons in terms of quarks bound by chiral fields require some justification. After all, by far the most successful models of hadron spectroscopy are constituent quark models, which consist of quarks and antiquarks interacting with confining forces [1, 2, 3, 4]. At short distances, the force reduces to a perturbative one-gluon exchange potential. At long distances, it is presumably governed by the string tension. However, models involving confining forces can deal only with a limited number of valence particles, in contrast to models which use relativistic fields to describe systems with infinite numbers of particles (filling the Dirac sea for example). They do not model the physical properties of the vacuum. This is also true of color-dielectric models which can only account for the dynamics of valence quarks. Constituent quark models require a roughly 300 MeV constituent quark mass as a phenomenological input, for the light u and d quarks.

The constituent quark mass is intimately linked to the spontaneously broken chiral symmetry of the physical vacuum, the existence of which is revealed by the exceptionally low mass of the $\pi\,(139)$. This is precisely what is modelled by quarks interacting with chiral fields. It is quite remarkable that the same chiral field which gives u and d quarks a roughly $300 - 400$ MeV mass can also bind the quarks so as to form hadrons of observed size and mass.

As long as chiral fields produce a bound state, it is probable that they are more relevant to the hadron structure than confining forces. However, the binding is usually too weak to allow for bound excited states. Confining forces begin to play a crucial role for radial excitations and for relative orbital angular momenta $L = 1$ or higher. In this book, we endeavor to show precisely where the confining forces begin to dominate over the forces produced by chiral fields. It is tantalizing to consider the wealth of spectroscopic data which lie at the intersection of these, so far incompatible domains.

This is a book about hadron models and their limitations. We intentionally omit discussing QCD (quantum chromodynamics). Too many contradictory models have been 'derived' from QCD. Our present understanding of low energy non-perturbative QCD is too patchy and it certainly does not show us how to model confinement and spontaneously broken chiral symmetry. With very few exceptions, such as instanton induced interactions, and in spite of innumerable claims to the contrary, the way various models relate to QCD remains largely unknown.

This is a frequent occurrence in physics and in no way does it disqualify the models. The nuclear shell model—which gives a quantitative unified description of all nuclei—can be derived from interactions between nucleons, which bear no resemblance to meson exchange potentials. Nuclear spectroscopy teaches us little about strong interactions, save for some of their symmetries. It is quite possible that we will need to expose several intermediate structures before being able to relate low energy hadron properties to QCD. Nor is there reason a priori to expect an imminent breakthrough. There has certainly been no lack of talented physicists during the 60 years which separated the explanation of superconductivity and its discovery early in this century.

Several reviews have been written of chiral models of hadrons, such as Skyrme and Nambu–Jona-Lasinio models. Rarely are the limitations of a model clearly spelled out, especially when it is presented as a natural outcome of QCD. In this book we attempt to show how the models relate, how they differ and where they fail. The author believes that the reader will learn more by being told about the calculations people do rather than what they say.

The book is addressed to the physicist who seeks an introduction to the vast literature on the subject and who might also want to make some calculations of his own. Familiarity with the Dirac equation and Lorentz covariance and some elementary knowledge of field theory are assumed. References are, however, given from which the basic techniques can be learnt.

The author has endeavored to quote useful references. The bibliography inevitably portrays a personal choice. In such a short book, the reader would rapidly become lost if all the published work were referred to.

Numerous people have helped the author understand various aspects of the models. He has greatly benefited from continued discussions with Ian Aitchison, Enrique Arriola, Roger Balian, Richard Ball, Mike Birse, Jean-Paul Blaizot, Wojtek Broniowski, Gerry Brown, Christo Christov, Mitya Diakonov, Michel Gaudin, Klaus Goeke, Fred Goldhaber, Jiři Hosek, Jörg Hüfner, Martine Jaminon, Sid Kahana, Ramon Mendez-Galain, Emil Nikolov, Vincent Pasquier, Victor Petrov, Pavel Pobylitsa, Michal Praszalowicz, Bruno Van den Bossche, and Wolfram Weise. The text does not, however, necessarily reflect their viewpoint.

Most of these discussions were made possible by support from the Oxford–Paris programme and an award from the Alexander von Humboldt Stiftung, without which the book would never have been completed.

The author is most indebted to E. Arriola, J. P. Blaizot, M. Jaminon and B. Van den Bossche, who were good enough to read parts of the manuscript and who made many comments which improved the text. The author did not agree with all the suggested changes and they should not be held responsible for whatever errors might remain.

The latex file was typed using the Scientific Workplace package.

G. R.

Paris
March 1997

CONTENTS

1

INTRODUCTION

The hadronic structure of the vacuum is governed by low-energy non-perturbative QCD and it determines the effective interactions of quarks, propagating in the vacuum. We have but scanty knowledge of the vacuum structure. We do, however, know the symmetries of the effective interactions, because they reflect the observed symmetries of the strong interactions (chapter 2). Symmetry is not sufficient to determine the effective quark interactions and additional assumptions are required to model them. Most of the effective interactions which have been used are inspired either by gluon exchange or by an assumed instanton structure of the vacuum. In some channels the two are similar but in other channels they differ significantly. They are described and compared in chapter 3.

Effective quark interactions are used to model the spontaneous chiral symmetry breaking in the vacuum, the mesons which are normal vibrational modes of the vacuum and the baryons which appear as bound states of quarks (solitons). The Nambu Jona-Lasinio model is a typical and generic example (chapter 4). Useful forms of the model are obtained by introducing auxiliary fields which represent quark bilinear operators, described in chapter 2. The scalar part of the auxiliary field is a chiral field so that the Nambu Jona-Lasinio model can also be viewed as a model describing quarks interacting with chiral fields. Most of the physics of the Nambu Jona-Lasinio model can be understood in terms of an approximate equivalent linear sigma model (chapter 5). In the mean field approximation, the model shows how quarks acquire a dynamical *constituent* mass, which is proportional to the vacuum expectation value of the scalar field. The constituent quark mass may be used as an order parameter to measure the spontaneous chiral symmetry breaking in the vacuum. In the chiral limit of vanishing *current* u and d quark masses, massless pions are generated as normal $(q\bar{q})$ excitation modes of the vacuum. In the other channels, mesons are generated with masses which are close their constituent quark masses. When u, d and s quarks are included, the mesons form scalar and pseudoscalar nonets which are discussed in sections 5.6.1 and 5.6.2.

In contrast to the linear sigma model, which can be renormalized, the Nambu Jona-Lasinio model needs to be regularized (chapter 6). The regularization is not a mathematical trick designed to bypass ultraviolet divergencies. It is a physical process the nature of which is embedded in the ill-understood structure of the vacuum. For example, when an instanton structure is assumed for the vacuum, the theory is regularized by the resulting non-locality of the effective quark interaction (section 3.2). The cut-off, which is introduced by the regularization has a low value, of the order of 1 GeV. This is its characteristic feature and it is dictated by phenomenology (section 5.5). As a result, calculated observables depend strongly on the value of the cut-off and sometimes also on the adopted regularization scheme. Each regularization scheme has some attractive features but they are all thrawn with problems. Non-local regularizations are only defined for euclidean 4-momenta thereby introducing a considerable arbitrariness in the calculation of on-shell masses. Local regularizations, such as the popular proper-time and Pauli–Villars regularizations, can only be applied to the real part of the action and they regularize only the fermion determinant. In addition, they introduce unphysical poles in the meson propagators. Regularizations limited to 3-momenta are cleaner because they correspond to a Hilbert space truncation. However, they break the Lorentz covariance of the model.

The full physical content of chiral quark models becomes apparent when they are used to calculate correlation functions. This is where, for example, one encounters the limitations due to their failure to confine quarks. The properties of correlation functions, together with the frequently used effective action formalism, are explained in chapter 7. When approximate evaluations of correlation functions are made, it is essential to preserve the symmetries of the underlying model so as to ensure, for example, that the pion remains a massless Goldstone boson in the chiral limit. This feature becomes critical for the calculation of contributions which are next to leading order in N_c. Two distinct symmetry conserving approximations are explained in chapter 8 . One is based on the effective action formalism and the other is based on a self-consistent use of the Dyson-Schwinger and Bethe–Salpeter equations. The symmetry conserving approximations are also expressed in terms of resummations of Feynman graphs. Calculations of the scalar and vector meson correlation functions in the Nambu Jona-Lasinio model are discussed in chapter 9. An overview of results obtained in the meson sector is given in chapter 10, where the theory is also extended to finite densities and temperatures.

Other chiral quark models, such as the Diakonov Petrov model, scaled models involving a dilaton field, the Gell-Mann Lévy model and color dielectric models, are compared in chapter 11. The renormalized Gell-Mann Lévy model gives rise to a low energy Landau ghost, which prevents the occurrence of a stable soliton (section 11.3.3).

The description of baryons in terms of chiral solitons is explained in chapter 12. Solitons arise as bound states of quarks when the chiral field acquires a hedgehog shape. The spectrum of quarks, propagating in a hedgehog field, allows one to understand the relation between the winding number of the hedgehog field and the baryon number of the soliton. Bound states of quarks are obtained when the proper-time regularization is used. However, ambiguities arise in the evaluation of observables because only the real part of the action is regularized. Radial stability of the proper-time regularized soliton is only ensured when the chiral field is restricted to the chiral circle, as in the Diakonov Petrov model.

Hedgehog fields yield solitons which have neither definite angular momentum nor isospin. Baryons with given angular momentum and flavor are described in terms of solitons which rotate in flavor space (chapter 13). The rotation modifies the expectation value of operators such as the axial constant g_A and form factors. A review of results obtained for the properties of the $N\,(939)$, $\Lambda\,(1116)$, $\Sigma\,(1194)$, and $\Xi\,(1318)$ octet is given. Taken together with meson properties, reviewed in chapter 10, one finds that the structure of the low lying mesons and baryons is well accounted for by quarks interacting with chiral fields.

2

QUARK BILINEAR OPERATORS

Bilinear forms of the quark field q, usually expressed in the form $\bar{q}\Gamma q$, are the building blocks of quark models. Gluon exchange interactions, instanton induced interactions (chapter 3) as well as the 4-fermion interactions of the Nambu Jona-Lasinio model (chapter 4) are all expressed in terms of such bilinear operators. Expectation values of bilinear operators $\bar{q}\Gamma q$ are quark condensates. Meson fields couple directly to bilinear operators which carry the same quantum numbers. They can be introduced into quark actions so as to linearize products of quark bilinears (section 4.3). In some cases, it is useful to eliminate completely the quark field from the effective action, in favor of the meson fields. So-called bosonized forms of effective actions are obtained this way. Bosonized effective actions involve derivatives of the meson fields to all orders. This reflects the fact that the meson field represents a composite $q\bar{q}$ system and that it has fewer degrees of freedom than the composite system. This is why Skyrme type models, which describe baryons in terms of meson fields with limited time derivatives, require to be amended by higher order derivative terms, such as the Wess–Zumino term.

2.1 Lorentz transformations of $\bar{q}\Gamma q$ bilinear operators

The bilinear operators $\bar{q}\Gamma q$ have specific transformation properties. Because $\bar{q}q$ is a scalar, the operators $\bar{q}\Gamma q$ bear the quantum numbers of the operator Γ. In diquark channels, the operators have the form $\bar{q}^c\Gamma q$ where q^c is the charge conjugate quark field. The behavior of quark bilinears under Lorentz transformations is well described in most field theory books (see [5] for example). In short, if Γ is an operator acting on the flavor and color indices, but *not* on the Dirac indices, then $\bar{q}\Gamma q$ is a Lorentz scalar (positive parity), $\bar{q}i\gamma_5\Gamma q$ is a pseudoscalar (negative parity), $\bar{q}\gamma_\mu\Gamma q$ is a vector (negative parity) and $\bar{q}\gamma_5\gamma_\mu\Gamma q$ an axial vector (positive parity). We shall always assign a phase to the quark bilinear operators so as to make them *hermitian*.

2.2 Isospin and flavor rotations, chiral and axial transformations

Quarks come in N_f flavors and N_c colors. The quark field q has N_f flavor and N_c color indices to distinguish them. Symmetry properties (see section 4.5) are expressed in terms of the following transformations of the quark fields.

In a *flavor rotation*, the quark fields transform as follows:

$$q \to e^{-i\theta_a\tau_a/2}q \qquad q^\dagger \to q^\dagger e^{i\theta_a\tau_a/2} \qquad \bar{q} \to \bar{q}e^{i\theta_a\tau_a/2} \tag{2.1}$$

Table 2.1 *Transformation properties of quark bilinears*

Operator	Flavor rotation	Chiral rotation	Axial transformation
$q \rightarrow$	$e^{-i\theta_a\tau_a/2}q$	$e^{-i\gamma_5\theta_a\tau_a/2}q$	$e^{-i\gamma_5\theta}q$
$\bar{q} \rightarrow$	$\bar{q}e^{i\theta_a\tau_a/2}$	$\bar{q}e^{-i\gamma_5\theta_a\tau_a/2}$	$\bar{q}e^{-i\gamma_5\theta}$
$\delta(\bar{q}q)$	0	$-\theta_a\bar{q}i\gamma_5\tau_a q$	$-2\theta\bar{q}i\gamma_5 q$
$\delta(\bar{q}\tau_a q)$	$-\theta_b f_{bac}\bar{q}\tau_c q$	$-\frac{2}{N_f}\theta_a\bar{q}i\gamma_5 q - \theta_b d_{bac}\bar{q}i\gamma_5\tau_c q$	$-2\theta\bar{q}i\gamma_5\tau_a q$
$\delta(\bar{q}i\gamma_5 q)$	0	$\theta_a\bar{q}\tau_a q$	$2\theta\bar{q}q$
$\delta(\bar{q}i\gamma_5\tau_a q)$	$-\theta_b f_{bac}\bar{q}i\gamma_5\tau_c q$	$\frac{2}{N_f}\theta_a\bar{q}q + \theta_b d_{bac}\bar{q}\tau_c q$	$2\theta\bar{q}\tau_a q$
$\delta(\bar{q}\gamma_\mu q)$	0	0	0
$\delta(\bar{q}\gamma_\mu\tau_a q)$	$-\theta_b f_{bac}\bar{q}\gamma_\mu\tau_c q$	$\theta_b f_{bac}\bar{q}\gamma_5\gamma_\mu\tau_c q$	0
$\delta(\bar{q}\gamma_5\gamma_\mu q)$	0	0	0
$\delta(\bar{q}\gamma_5\gamma_\mu\tau_a q)$	$-\theta_b f_{bac}\bar{q}\gamma_5\gamma_\mu\tau_c q$	$\theta_b f_{bac}\bar{q}\gamma_\mu\tau_c q$	0

where the τ_a are the $N_f^2 - 1$ hermitian traceless matrices of an $SU(N_f)$ Lie algebra. They act on the *flavor variables* of the quark fields. Their properties are given in appendix A. The parameters θ_a are the amplitudes of the flavor rotation. Flavor rotations form an $SU(N_f)$ group.

Isospin rotations form an $SU(2)$ subgroup of the $SU(N_f)$ group:

$$q \rightarrow e^{-i\sum_{a=1}^{3}\theta_a\tau_a/2}q \qquad q^\dagger \rightarrow q^\dagger e^{i\sum_{a=1}^{3}\theta_a\tau_a/2} \qquad \bar{q} \rightarrow \bar{q}e^{i\sum_{a=1}^{3}\theta_a\tau_a/2} \tag{2.2}$$

An isospin rotation involves only the 3 generators $\tau_{a=1,2,3}$ which have the same commutation relations as the Pauli matrices, generators of the $SU(2)$ group. Strong interactions are invariant under isospin rotations provided that the current masses of the u and d quarks are taken to be equal.

In a *chiral transformation*, the quark fields transform as follows:

$$q \rightarrow e^{-i\gamma_5\theta_a\tau_a/2}q \qquad q^\dagger \rightarrow q^\dagger e^{i\gamma_5\theta_a\tau_a/2} \qquad \bar{q} \rightarrow \bar{q}e^{-i\gamma_5\theta_a\tau_a/2} \tag{2.3}$$

Although chiral transformations do not form a group (see section 2.3), they are often improperly referred to as *axial* $SU(N_f)$ transformations.

In an *axial transformation*, the quark fields transform as follows:

$$q \rightarrow e^{-i\gamma_5\theta}q \qquad q^\dagger \rightarrow q^\dagger e^{i\gamma_5\theta} \qquad \bar{q} \rightarrow \bar{q}e^{-i\gamma_5\theta} \tag{2.4}$$

The transformation (2.4) is often referred to as an axial $U_A(1)$ transformation.

Table 2.1 gives the transformation properties of various quark bilinears under the corresponding *infinitesimal* transformations:

For example, in an infinitesimal chiral rotation, the operator $\bar{q}\tau_a q$ transforms as:

$$\bar{q}\tau_a q \rightarrow \bar{q}\tau_a q - \frac{2}{N_f}\theta_a\bar{q}i\gamma_5 q - \theta_b d_{bac}\bar{q}i\gamma_5\tau_c q \tag{2.5}$$

The coefficients f_{abc} and d_{abc} are defined in section A.1. Because f_{abc} is antisymmetric, the operators $(\bar{q}\tau_a q)^2, (\bar{q}i\gamma_5\tau_a q)^2, (\bar{q}\gamma^\mu\tau_a q)(\bar{q}\gamma_\mu\tau_a q)$ ... are invariant under flavor rotations. Chiral transformations leave the following operators invariant:

$$\frac{2}{N_f}(\bar{q}q)^2 + (\bar{q}\tau_a q)^2 + \frac{2}{N_f}(\bar{q}i\gamma_5 q)^2 + (\bar{q}i\gamma_5\tau_a q)^2 \equiv \sum_{a=0}^{N_f^2-1}\left[(\bar{q}\tau_a q)^2 + (\bar{q}i\gamma_5\tau_a q)^2\right]$$

$$(\bar{q}\gamma^\mu\tau_a q)(\bar{q}\gamma_\mu\tau_a q) + (\bar{q}\gamma_5\gamma^\mu\tau_a q)(\bar{q}\gamma_5\gamma_\mu\tau_a q)$$

$$(\bar{q}\gamma^\mu q)(\bar{q}\gamma_\mu q) \qquad (\bar{q}\gamma_5\gamma^\mu q)(\bar{q}\gamma_5\gamma_\mu q) \tag{2.6}$$

In (2.6) we have defined the matrix $\tau_{a=0}$ to be proportional to the unit matrix:

$$\tau_{a=0} = \sqrt{\frac{2}{N_f}}I \tag{2.7}$$

When $N_f = 2$, we have $d_{abc} = 0$ and chiral transformations also leave the following separate operators invariant:

$$(\bar{q}q)^2 + (\bar{q}i\gamma_5\tau_a q)^2 \qquad (\bar{q}\tau_a q)^2 + (\bar{q}i\gamma_5 q)^2 \qquad (N_f = 2) \tag{2.8}$$

The two sets of operators transform into each other in axial transformations.

2.3 Right and left rotations

Flavor rotations form a $SU(N_f)$ group but chiral transformations do not. This is why it is more proper to define right and left rotations, each of which forms an $SU(N_f)$ group. We first define right and left quark fields:

$$q_i^R = \frac{1+\gamma_5}{2}q_i \qquad q_i^L = \frac{1-\gamma_5}{2}q_i \qquad q_i = q_i^R + q_i^L$$

$$\bar{q}_i^R = \bar{q}_i\frac{1-\gamma_5}{2} \qquad \bar{q}_i^L = \bar{q}_i\frac{1+\gamma_5}{2} \qquad \bar{q}_i = \bar{q}_i^R + \bar{q}_i^L$$

$$\bar{q}_i^R q_j^R = \bar{q}_i^L q_j^L = 0 \qquad \bar{q}_i q_j = \bar{q}_i^R q_j^L + \bar{q}_i^L q_j^R$$

$$\bar{q}_i(i\partial_\mu\gamma^\mu)q_j = \bar{q}_i^R(i\partial_\mu\gamma^\mu)q_j^R + \bar{q}_i^L(i\partial_\mu\gamma^\mu)q_j^L \tag{2.9}$$

In the definitions (2.9), the subscripts i and j denote (color, flavor,...) indices *other* than Dirac indices.

A *right flavor rotation* is defined to be a flavor rotation of the right quark field only:

$$q^R \to e^{-i\theta_a\tau_a/2}q^R \qquad \bar{q}^R \to \bar{q}^R e^{i\theta_a\tau_a/2} \qquad q^L \to q^L \qquad \bar{q}^L \to \bar{q}^L \tag{2.10}$$

Table 2.2 *Quantum numbers of colorless $\bar{q}\Gamma q$ operators and associated flavor nonets*

$\bar{q}\Gamma q$	$^{2S+1}L_J$	J^{PC}	u, d, s nonet (MeV)
$\bar{q}i\gamma_5\tau_a q$	$^1S_0$	0^{-+}	$\pi\,(138)\,, \eta\,(547)\,, \eta'\,(958)\,, K\,(494)$
$\bar{q}\gamma_\mu\tau_a q$	$^3S_1$	1^{--}	$\rho\,(770)\,, \omega\,(782)\,, \phi\,(1020)\,, K^*\,(892)$
$\bar{q}\partial_\mu\gamma_5\tau_a q$	$^1P_1$	1^{+-}	$b_1\,(1235)\,, h_1\,(1170)\,, h_1\,(1380)\,, K_{1B}$
$\bar{q}\tau_a q$	$^3P_0$	0^{++}	$a_0\,(980)\,, f_0\,(1370)\,, f_0\,(980)\,, K_0^*\,(1430)$
$\bar{q}\gamma_5\gamma_\nu\tau_a q$	$^3P_1$	1^{++}	$a_1\,(1260)\,, f_1\,(1285)\,, f_1\,(1510)\,, K_{1A}$
$i\bar{q}\left(\gamma_\mu\partial_\nu + \gamma_\nu\partial_\mu + \frac{2}{3}\eta_{\mu\nu}\partial\!\!\!/\right)\tau_a q$	$^3P_2$	2^{++}	$a_2\,(1320)\,, f_2\,(1270)\,, f_2'\,(1525)\,, K_2^*\,(1430)$

A *left flavor rotation* is defined to be a flavor rotation of the left field only:

$$q^R \to q^R \qquad \bar{q}^R \to \bar{q}^R \qquad q^L \to e^{-i\theta_a\tau_a/2}q^L \qquad \bar{q}^L \to \bar{q}^L e^{i\theta_a\tau_a/2} \tag{2.11}$$

In a flavor rotation, the left and right fields q^L and q^R both undergo the same rotation. However, in a chiral rotation, the left and right fields undergo inverse rotations.

Clearly, right and left rotations each form an $SU(N_f)$ group. Together, right and left transformations form the $SU_R(N_f) \otimes SU_L(N_f)$ group. If an operator is invariant under left and right rotations, it is also invariant under flavor and chiral transformations. In spite of being less proper, it is often more convenient to work with the flavor and chiral transformations than with right and left transformations. The same holds for the axial transformations, which do not form a group.

Transformations acting on color indices of the quark field can be defined in the same way as those acting on flavor variables. They form the $SU(N_c)$ group.

2.4 Quantum numbers of $\bar{q}\Gamma q$ operators

Mesons are quark–antiquark excitations of the vacuum, usually referred to as $q\bar{q}$ excitations. They can be created by applying to the vacuum bilinear operators of the form $\bar{q}\Gamma q$, in which the Γ are operators which can act on the Dirac, flavor and color variables of the quark field. The operator Γ determines the quantum numbers (spin, parity, isospin, charge conjugation,...) of the bilinear operator $\bar{q}\Gamma q$.

When the matrix operator Γ does not act on the color variables of the quark field, the bilinear $\bar{q}\Gamma q$ is *colorless* (a scalar with respect to color rotations). Examples are listed in Table 2.2 which shows the quantum numbers of various colorless $\bar{q}\Gamma q$ operators as well as the associated $U(3)_f$ flavor nonets which they can excite :

In Table 2.2, $\bar{q}(x)\,\eta_{\mu\nu}\Gamma q(x) \equiv \left(\frac{\partial_\mu\partial_\nu}{\partial^2} - g_{\mu\nu}\right)\bar{q}(x)\,\Gamma q(x)$. The phases are chosen so as to make the bilinear operators $\bar{q}\Gamma q$ *hermitian*. Further examples are listed in [6]. The τ_a are the N_f^2 matrices of the $U(N_f)$ group, defined in (A.22). The meson masses (in parentheses) are expressed in MeV. The K_{1A} and K_{1B} are nearly 45° mixed states of $K_1(1270)$ and $K_1(1400)$. Data are taken from the 1994 Particle Data [7]. In the 1996 Particle Data [8], the assignments of several mesons belonging to the nonet in the scalar channel $\bar{q}\tau_a q$ are questioned. This is further discussed in section 5.6.2. The operators $\bar{q}\Gamma q$ listed above are not unique. For example, the operators $\partial_\mu \bar{q}\gamma_5\tau_a q$ and $\bar{q}\gamma_\mu\gamma_5\tau_a q$ can both excite the a_1 meson. An operator is usually chosen so as to have a strong overlap with the meson under study. Further bilinear forms are listed in [9, 10] and the forms for arbitrary spin mesons are discussed in [11, 12]. One way to check the quantum numbers in Table 2.2 is to apply the operator $\bar{q}\lambda_c\Gamma q$ to a Dirac sea $|D\rangle$. For example, a $q\bar{q}$ excitation with zero momentum, spin $S = 1$, total angular momentum $J = 1$ and positive parity can be constructed as follows:

$$\int d_3r\ \bar{q}(\boldsymbol{r})\gamma_5\gamma q(\boldsymbol{r})|D\rangle = \sum_{\boldsymbol{k}\sigma_1\sigma_2} \left\langle \sigma_1 \left| i\frac{\boldsymbol{k}\times\vec{\sigma}}{m} \right| \sigma_2 \right\rangle a^\dagger_{\boldsymbol{k}\sigma_1} a_{\boldsymbol{k}\sigma_2} |D\rangle \tag{2.12}$$

where the operator $a^\dagger_{\boldsymbol{k}\sigma}$ creates a quark with momentum $\boldsymbol{k}$ and spin projection σ.

Bilinear operators of the form $\bar{q}\lambda_c\Gamma q$ are *color octets*. In these forms, the λ_c are the $SU(N_c)$ Gell-Mann matrices, which act on the *color* variables of the quark fields (not to be confused with the τ_a in Table 2.2 which act on the flavor variables). A common example is the color current operator $\bar{q}\gamma_\mu\lambda_c q$ to which the gluons couple in QCD and which is used to construct gluon exchange potentials (see section 3.1).

2.5 Diquark bilinear operators

In diquark channels, it is useful to introduce the charge conjugate quark field:

$$q_i^c = C_{ij}\bar{q}_j \qquad q_i C_{ij} = -\overline{q^c}_j \qquad C^*_{ik}\bar{q}_k = -q_i^c$$

$$C = i\gamma^2\gamma^0 = -C^{-1} = -\widetilde{C} = -C^\dagger \tag{2.13}$$

where C is the charge conjugation matrix which acts on the Dirac indices of the quark fields [5]. One should bear in mind that charge conjugate fields have opposite baryon number. Thus the operator q^c adds one quark to the system and $\overline{q^c}$ removes one. Bilinear operators are then constructed of the form $\overline{q^c}\Gamma q$ or $\bar{q}\Gamma q^c$ which have respectively baryon number $-2/3$ and $2/3$.

For $N_c = 3$, the fields $\overline{q^c}$ and q are color triplets. A diquark state can then be either an antisymmetric *color triplet* state, which is excited by the three operators $\overline{q^c}\lambda^A\Gamma q$, or a symmetric *color sextet* state, which is excited by the six operators $\overline{q^c}\lambda^S\Gamma q$, where the operator Γ is assumed not to act on the color variables. The six symmetric matrices are $\lambda^S = \left(\sqrt{\frac{2}{3}}I, \lambda_1, \lambda_3, \lambda_4, \lambda_6, \lambda_8\right)$ and

Table 2.3 *Color triplet diquark operators*

$\bar{q}\lambda^A\Gamma q^c$	$^{2S+1}L_J$	Parity	Flavor
$\bar{q}\lambda^A i\gamma_5\tau^{A'} q^c$	$^1S_0$	$+$	$\bar{3}$
$\bar{q}\lambda^A\tau^{A'} q^c$	$^3P_0$	$-$	$\bar{3}$
$\bar{q}\lambda^A\gamma_\mu\tau^S q^c$	$^3S_1$	$+$	6
$\bar{q}\lambda^A\gamma^\mu\gamma_5\tau^A q^c$	$^3P_1$	$-$	$\bar{3}$

the three antisymmetric matrices are $\lambda^A = (\lambda_2, \lambda_5, \lambda_7)$. The triplet and sextet diquark operators correspond to the $SU(3)_c$ addition rule: $3 \otimes 3 = \bar{3} + 6$. A color triplet ($\bar{3}$) diquark state can couple to a color triplet (3) single quark state to form a color singlet. This is why the color triplet ($\bar{3}$) diquark bilinears are used to study colorless baryons in terms of quark–diquark substructures [13, 14, 15, 16]. A color triplet diquark state is antisymmetric with respect to the exchange of the color indices of the fields. Color triplet bilinears can therefore also be written in the form $\epsilon_{\alpha\beta\gamma}\left(\overline{q^c}\right)_\beta \Gamma q_\gamma$ where the indices (α, β, γ) are color indices of the quark fields.

With u, d and s quarks, one can construct *flavor triplet* and *flavor sextet* diquark operators, according to the $SU(3)_f$ addition rule of flavors: $3 \otimes 3 = \bar{3} + 6$. For color triplets, the flavor triplet and sextet bilinears have the form $\epsilon_{\alpha\beta\gamma}\left(\overline{q^c}\right)_\beta \tau^A \Gamma q_\gamma$ and $\epsilon_{\alpha\beta\gamma}\left(\overline{q^c}\right)_\beta \tau^S \Gamma q_\gamma$ respectively, where the matrices τ^A and τ^S act on the flavor variables of the quark fields. The six symmetric and three antisymmetric flavor matrices τ^S and τ^A are defined in the same way as the corresponding color matrices.

A color triplet diquark state is antisymmetric in the exchange of the colors. According to the Pauli principle, it must be symmetric with respect to the exchange of the other coordinates. The exchange of spatial and spin coordinates yields a factor $(-)^{L+S+1}$. The flavor wavefunction must therefore be odd $\left(\tau^A\right)$ if $L + S$ is even and even $\left(\tau^S\right)$ when $L + S$ is odd. Some color triplet diquark operators are listed in Table 2.3

The phases are chosen such that hermitian conjugate of $\bar{q}^c\lambda^A\Gamma q$ is $\bar{q}\lambda^A\Gamma q^c$.

3

EFFECTIVE QUARK INTERACTIONS

Quark models are usually expressed in terms of effective interactions, which reflect the symmetries of strong interactions. However, symmetry is not sufficient to determine them. The forms of the effective interactions are further dictated by their phenomenological fitness and by their simplicity of use. We should keep in mind that we are not discussing the effective interactions with the aim of adding them to a confining potential. Rather, we have the intention of using them in models which can account for the spontaneously broken chiral symmetry of the vacuum and, hopefully also, for the formation of hadrons. Attempts to derive effective quark interactions from QCD have followed different directions. Some attempts, sometimes referred to as the global color model, lead to an effective gluon exchange interaction, the space–time dependence of which is governed by a phenomenologically determined effective gluon propagator [17, 18, 19, 20, 21, 22] or which is based on the so-called field strength approach [23]. Other attempts, based on so-called hadronization methods, introduce auxiliary non-local fields which re-sum ladders of gluon exchange diagrams [14, 24]. Another direction, which derives a different effective interaction between quarks, stems from a study of their propagation in an instanton liquid, which describes the QCD vacuum [25, 26, 27]. In this chapter we describe the various interactions which are commonly used and we do not attempt to derive them from QCD.

3.1 Gluon exchange interactions

Gluon exchange interactions, with phenomenologically modified gluon propagators, are very often used to generate the effective interactions between quarks. In the large N_c limit, Fierz transformations can be used to express them in terms of colorless quark bilinear forms. This leads to effective interactions between quarks which can generate spontaneous chiral symmetry breaking in the vacuum. However, one should bear in mind that there is no a priori reason to expect that the effective interaction, which arises from the little understood low energy QCD, should have the form of a gluon exchange interaction. For example, there is evidence that gluon exchange interactions are less important than the exchange of colorless pseudoscalar mesons (the Goldstone bosons) in the *residual* interactions between quarks in confining constituent quark models [4, 28].

The exchange of a gluon between quarks contributes an interaction term to the effective action, which has the form:

$$I_{int}\left(q,q^{\dagger}\right)=-\frac{\alpha^{2}}{2}\int d_{4}xd_{4}y\left[\bar{q}\left(x\right)\gamma^{\mu}\lambda_{c}q\left(x\right)\right]D_{\mu\nu}\left(x,y\right)\left[\bar{q}\left(y\right)\gamma^{\nu}\lambda_{c}q\left(y\right)\right] \tag{3.1}$$

The factors $\bar{q}\gamma^{\mu}\lambda_c q$ are the color octet quark currents, $D_{\mu\nu}(x,y)$ is the gluon propagator expressed in some gauge and α^2 is a quark–gluon coupling constant. Gluon exchange dominates when a high space-like momentum is exchanged. However, at low momentum transfer between the quarks, the vacuum polarization makes the perturbative running coupling constant very large. Models which lead to gluon exchange interactions assume that the strong non-perturbative effects can be taken into account by modifying the gluon propagator [21, 22]. It is usually modified so as to suit phenomenology. The gluon propagator could also be chosen so as to yield a confining potential. However, this won't do for our purpose, because we do not know how to calculate spontaneous chiral symmetry breaking with a confining interaction. Although there is no a priori reason to expect that the effective interaction of quarks propagating in the QCD vacuum should retain the form of a gluon exchange interaction, we shall see that the gluon exchange potential has several features which point in the right direction. The effective gluon propagator also provides a natural way to introduce a non-locality into the quark–quark interaction. This non-locality can be used to regularize the ultraviolet divergences of the model [14, 20] (see sections 3.4 and 6.1).

A useful starting point for low energy effective actions consists in using a gluon propagator which is reduced to its crudest form:

$$D_{\mu\nu}(x,y) \approx \frac{1}{\Lambda^2}\delta(x-y)\, g_{\mu\nu} \tag{3.2}$$

where Λ^2 is a constant with the dimension E^2 of an energy squared. This leads to a local interaction of the form:

$$\mathcal{L}_{int}(x) = -\frac{\alpha^2}{2\Lambda^2}\left[\bar{q}(x)\gamma^{\mu}\lambda_c q(x)\right]\left[\bar{q}(x)\gamma_{\mu}\lambda_c q(x)\right] \tag{3.3}$$

The local form (3.3) does not, of course, regularize the ultraviolet divergences.

3.1.1 *Gluon exchange in the colorless $q\bar{q}$ channel*

Fierz transforms can be used to express the gluon exchange interaction (3.3) in the other $q\bar{q}$ channel (see appendix A). Using (A.11), (A.4) and (A.24) for the color, flavor and Dirac indices respectively, and taking also into account the minus sign which is due to an odd permutation of the quark fields, we can write the gluon exchange interaction (3.3) in the form:

$$\mathcal{L}_{int} = \frac{\alpha^2}{2\Lambda^2}\frac{\left(N_c^2-1\right)}{N_c^2}\sum_{a=0}^{N_f^2-1}\Big[\left(\bar{q}\tau_a q\right)\left(\bar{q}\tau_a q\right) + \left(\bar{q}i\gamma_5\tau_a q\right)\left(\bar{q}i\gamma_5\tau_a q\right)$$

$$-\frac{1}{2}\left(\bar{q}\gamma^{\mu}\tau_a q\right)\left(\bar{q}\gamma_{\mu}\tau_a q\right) - \frac{1}{2}\left(\bar{q}\gamma_5\gamma^{\mu}\tau_a q\right)\left(\bar{q}\gamma_5\gamma_{\mu}\tau_a q\right)\Big]$$

$$-\frac{\alpha^2}{2\Lambda^2}\frac{1}{2N_c}\sum_{a=0}^{N_f^2-1}\sum_{c=1}^{N_c^2-1}\Big[(\bar{q}\tau_a\lambda_c q)(\bar{q}\tau_a\lambda_c q)+(\bar{q}i\gamma_5\tau_a\lambda_c q)(\bar{q}i\gamma_5\tau_a\lambda_c q)$$

$$-\frac{1}{2}(\bar{q}\gamma^\mu\tau_a\lambda_c q)(\bar{q}\gamma_\mu\tau_a\lambda_c q)-\frac{1}{2}(\bar{q}\gamma_5\gamma^\mu\tau_a\lambda_c q)(\bar{q}\gamma_5\gamma_\mu\tau_a\lambda_c q)\Big] \quad (3.4)$$

The first group of terms involves *colorless* $\bar{q}\Gamma q$ bilinear operators and the second group involves *color octet* bilinear operators of the form $\bar{q}\Gamma\lambda_c q$ (see section 2.4). When colorless quark condensates, solitons and meson propagators are calculated, the second term of (3.4) can be neglected in the large N_c limit.

From (2.6) we see that both the gluon exchange interaction (3.3) and its Fierz transform (3.4) *are invariant under flavor and chiral transformations* (2.1) and (2.3). They are also invariant under axial $U_A(1)$ transformations (2.4). In fact, the first term of (3.4) has the same form as the four-fermion interaction (4.1) used in Nambu Jona-Lasinio models (see chapter 4). The simplified gluon exchange interaction (3.4) relates the scalar and vector interaction strengths g_s^2 and g_v^2 of the Nambu Jona-Lasinio model to the quark–gluon effective coupling constant [29]:

$$g_s^2=\frac{8}{9}\frac{\alpha^2}{\Lambda^2} \qquad g_v^2=\frac{4}{9}\frac{\alpha^2}{\Lambda^2} \quad (3.5)$$

For the purpose of N_c counting, the bare quark–gluon coupling constant α^2 is assumed to be proportional to $1/N_c$. This is assumed to remain true for the coupling constants (3.5) of the effective interaction (3.4). The gluon exchange interaction (3.4) gives the right sign for the scalar and vector interaction strengths. The signs are significant. The scalar interaction needs to be attractive if it is to produce quark condensation and spontaneous chiral symmetry breaking. A repulsive time component $\frac{1}{2}(q^\dagger\tau_a q)(q^\dagger\tau_a q)$of the vector interaction is needed to produce a Coulomb repulsion and an attractive interaction $-\frac{1}{2}(\bar{q}\boldsymbol{\gamma}\tau_a q)\cdot(\bar{q}\boldsymbol{\gamma}\tau_a q)$ is required to bind $q\bar{q}$ pairs so as to form ω and ρ mesons. A fit to the colorless pseudoscalar and vector nonets, as quoted in [29], yields $g_s^2/g_v^2=1.5$. A fit to the $\rho(770)$ in [30] yields $g_s^2/g_v^2=1.34$. See also Table 10.3. This is only in qualitative agreement with the gluon exchange potential, which predicts a ratio $g_s^2/g_v^2=2$. If we assume a constant $\Lambda=0.91$ GeV in the effective gluon propagator (3.2), of the same order of magnitude as the cut-off required in the Nambu Jona-Lasinio model (see Table 10.1), then effective quark–gluon coupling constants $\frac{\alpha^2}{4\pi}=1.42$ and 1.88 are deduced from the scalar and vector interaction strengths [29]. The flavor independent gluon-exchange interaction predicts equal coupling constants for the vector and axial-vector interactions and equal interaction strengths for the isoscalar and isovector vector interactions. The latter leads to equal masses for the near degenerate $\rho(770)$ and $\omega(782)$ mesons. The fact that the simplified gluon exchange interaction has features which are well suited for phenomenological models is worth noting. Of course, the instantaneous approximation of the gluon propagator ruins the confining property of the $q\bar{q}$ interaction.

Nambu Jona-Lasinio models usually neglect the second term in (3.4), which is formed by color octet bilinears $\bar{q}\Gamma\lambda_c q$. In leading order in N_c they do not contribute to the constituent quark mass and to the colorless meson propagators. They would however contribute to the next to leading order effects such as meson loop corrections, which represent the quantum fluctuations of the vacuum.

The effective gluon propagator will in general be non-local [19]. A possible, although admittedly gauge-dependent, form could be:

$$D_{\mu\nu}(x,y) = a g_{\mu\nu} \langle x|r|y\rangle \tag{3.6}$$

where the operator r is diagonal in momentum space: $\langle k|r|k'\rangle = \delta_{kk'} r\left(k^2\right)$ (see section 3.4). The function $r\left(k^2\right)$ is a decreasing function of the euclidean momentum. In this case, the Fierz transforms can still be applied and the local interaction (3.4) is then replaced by a non-local interaction, which contributes a term of the form:

$$I_{int}\left(q,q^{\dagger}\right) = a\alpha^2 \frac{\left(N_c^2-1\right)}{N_c^2} \int d_4x d_4y \left[\left(\bar{q}(x)\tau_a q(y)\right)_x \langle x|r|y\rangle \left(\bar{q}(y)\tau_a q(x)\right)_y + \ldots \right] \tag{3.7}$$

to the *euclidean* lagrangian (defined in section 4.3). The non-locality can regularize the theory in the euclidean region (see section 6.1). For large euclidean momenta, the function $r(k)$ can be calculated in perturbative QCD, where it is found to decrease as $1/k^2$ [22]. Models which are regularized by non-local interactions are well suited to euclidean dynamics but it is far from clear how to use them to calculate, for example, on-shell meson masses, because such calculations require an analytic continuation of the regulator $r\left(k^2\right)$ from space-like $k^2 > 0$ to time-like $k^2 < 0$ momenta (see section 7.2.1).

3.1.2 *Gluon exchange in the diquark channel*

If we wish to study the quark–diquark structure of baryons, we can use the second Fierz transforms (see appendix A) in order to express the gluon exchange interaction in terms of the diquark operators, which are defined in section 2.5. Using (A.16), (A.4) and (A.30) for the color, flavor and Dirac indices respectively, the gluon exchange interaction (3.3) can be written in the form:

$$\mathcal{L}_{int} = -\frac{\alpha^2}{2\Lambda^2}\left(\bar{q}\gamma^{\mu}\lambda_c q\right)\left(\bar{q}\gamma_{\mu}\lambda_c q\right)$$

$$= \frac{\alpha^2}{2\Lambda^2}\left(1+\frac{1}{N_c}\right)\frac{1}{2}\left[\left(\bar{q}\tau^A\lambda^{A'}q^c\right)\left(\overline{q^c}\tau^A\lambda^{A'}q\right) + \left(\bar{q}\tau^A\lambda^{A'}i\gamma_5 q^c\right)\left(\overline{q^c}\tau^A\lambda^{A'}i\gamma_5 q\right)\right.$$

$$\left. -\frac{1}{2}\left(\bar{q}\tau^S\lambda^A\gamma^{\mu}q^c\right)\left(\overline{q^c}\tau^S\lambda^A\gamma_{\mu}q\right) + \frac{1}{2}\left(\bar{q}\tau^A\lambda^{A'}\gamma^{\mu}\gamma_5 q^c\right)\left(\overline{q^c}\tau^A\lambda^{A'}\gamma_{\mu}\gamma_5 q\right)\right]$$

$$-\frac{\alpha^2}{2\Lambda^2}\left(1-\frac{1}{N_c}\right)\frac{1}{2}\left[\left(\bar{q}\tau^S\lambda^{S'}q^c\right)\left(\overline{q^c}\tau^S\lambda^{S'}q\right)+\left(\bar{q}\tau^S\lambda^{S'}i\gamma_5 q^c\right)\left(\overline{q^c}\tau^S\lambda^{S'}i\gamma_5 q\right)\right.$$

$$\left.-\frac{1}{2}\left(\bar{q}\tau^A\lambda^S\gamma^\mu q^c\right)\left(\overline{q^c}\tau^A\lambda^S\gamma_\mu q\right)+\frac{1}{2}\left(\bar{q}\tau^S\lambda^{S'}\gamma^\mu\gamma_5 q^c\right)\left(\overline{q^c}\tau^S\lambda^{S'}\gamma_\mu\gamma_5 q\right)\right] \tag{3.8}$$

where $(q^c)_i = C_{ij}(\bar{q})_j$ is the charge conjugate field (2.13) and where the τ^S, λ^S and τ^A, λ^A are the symmetric and antisymmetric matrices acting respectively on the flavor and color indices of the quark fields. The first and second terms involve respectively color triplet and color sextet diquarks. In deriving the expression (3.8) we have used the fact that, for color triplet states, the flavor wavefunction must be even (τ^S) when $L+S$ is odd and vice versa, as shown and discussed in Table 2.3. For color sextet states, it is the other way around.

The gluon exchange interaction produces an attraction in the diquark color triplet channel. This is the basis for the description of baryons in terms of quark–diquark structures. Baryons are formed of a bound diquark in a color triplet $\bar{3}$ state which is coupled to a quark in a color triplet 3 state [14, 15].

3.2 Instanton induced quark interactions

Spontaneous chiral symmetry breaking in the vacuum can also result from instanton induced interactions. In [26, 27, 31] it is shown that a liquid consisting of N instantons and N anti-instantons in a 4-dimensional volume $V = \beta\Omega$ induces an interaction term to the *euclidean* effective action of the form:

$$I_{int}\left(q, q^\dagger\right) = -\left(Y_{N_f}^+ + Y_{N_f}^-\right) \qquad Y_{N_f}^\pm = \left(\frac{2V}{N}\right)^{N_f-1}\int d_4x \det_{N_f} J^\pm(x) \tag{3.9}$$

where N_f is the number of quark flavors. The use of a euclidean action is explained in section 4.3. In the expression (3.9), $J^\pm(x)$ is the $N_f \times N_f$ matrix:

$$J_{ij}^\pm(x) = \int d_4y d_4z\, \bar{q}_j(y)\left\langle y\left|\sqrt{M}\right|x\right\rangle\frac{1\pm\gamma_5}{2}\left\langle x\left|\sqrt{M}\right|z\right\rangle q_i(z) \tag{3.10}$$

and $q_i(z)$ is the quark field with *flavor* index i.

The operator M has dimension of a mass (energy) and it is diagonal in momentum space:

$$\langle k|M|k'\rangle = \delta_{kk'}M_0\left[r(k\rho/2)\right]^2 \tag{3.11}$$

The 'regulator' $r(z)$ is equal to:

$$r(z) = -z\frac{d}{dz}\left(I_0(z)K_0(z) - I_1(z)K_1(z)\right)$$

$$r(z=0)=1 \qquad r(z) \underset{z\to\infty}{\to} \frac{3}{4z^3} \tag{3.12}$$

where ρ is the instanton size. The operator M plays the role of a momentum-dependent constituent quark mass. It effectively regulates the ultraviolet divergences (see chapter 6).

We can define *delocalized quark fields* $\psi(x)$ and $\bar{\psi}(x)$ as follows:

$$\psi(x)=\int d_4y\,\langle x\,|r|\,y\rangle\, q(y) \qquad \bar{\psi}(x)=\int d_4y\bar{q}(y)\,\langle y\,|r|\,x\rangle \tag{3.13}$$

The fields $\psi(x)$ and $q(x)$ have the same dimensions and $\psi(x)$ reduces to $q(x)$ when $r=1$. We then have the more compact expression:

$$J_{ij}^{\pm}(x)=M_0\bar{\psi}_j(x)\,\frac{1\pm\gamma_5}{2}\psi_i(x)\equiv M_0K_{ij}^{\pm}(x) \tag{3.14}$$

and the instanton induced interaction between the quarks acquires the form of a 't Hooft determinant:

$$I_{int}\left(\psi,\psi^{\dagger}\right)=-\frac{N}{2V}\left(\frac{2M_0V}{N}\right)^{N_f}\int d_4x\left(\det K^{+}+\det K^{-}\right) \tag{3.15}$$

Note the dimensions: $\det K$ has dimension E^{3N_f}, $\left(\frac{2M_0V}{N}\right)^{N_f}$ has dimension E^{-3N_f} and N/V has dimension E^4. Thus, with d_4x of dimension E^{-4}, the interaction term $I\left(\psi,\psi^{\dagger}\right)$ is dimensionless. The instanton density N/V is proportional to N_c, so that the instanton induced interaction (3.15) is proportional to $(N_c)^{1-N_f}$. The expectation value of $\det K^{\pm}$ is proportional to $(N_c)^{N_f}$, so that the action is proportional to N_c. This ensures, for example, that solitons calculated with the interaction (3.15) have energies proportional to N_c and that meson masses are proportional to N_c^0.

The matrices $K^{\pm}$ can be written in the form:

$$K_{ij}^{+}=\bar{\psi}_j^L\psi_i^R=\bar{\psi}_j\frac{1+\gamma_5}{2}\psi_i \qquad K_{ij}^{-}=\bar{\psi}_j^R\psi_i^L=\bar{\psi}_j\frac{1-\gamma_5}{2}\psi_i \tag{3.16}$$

In a right flavor rotation, defined by (2.10), the matrices $K^{\pm}$ transform as:

$$K^{+}\to e^{-i\theta_a\tau_a/2}K^{+} \qquad K^{-}\to K^{-}e^{i\theta_a\tau_a/2} \tag{3.17}$$

A right transformation leaves $(\det K^{+}+\det K^{-})$ invariant. The same holds for left rotations. The instanton induced interaction is thus invariant under both flavor and chiral rotations.

In a $U_A(1)$ axial transformation, defined by (2.4), the matrices K^+ and K^- transform as:

$$K^+ \to e^{-i\theta} K^+ e^{-i\theta} \qquad K^- \to e^{i\theta} K^- e^{i\theta} \tag{3.18}$$

In the axial transformation, we therefore have:

$$\det K^+ + \det K^- \to e^{-2i\theta N_f} \det K^+ + e^{2i\theta N_f} \det K^- \tag{3.19}$$

The instanton induced interaction, which has the *form* of a 't Hooft determinant [32], is thus not invariant under axial transformations. It solves the $U_A(1)$ problem in the sense that it yields a mass for the $\eta'(958)$, thereby avoiding the necessity of adding extra phenomenological $U_A(1)$ symmetry breaking terms (see section 4.5.3). Calculations of the pseudoscalar nonet using the instanton induced interaction are discussed in [33].

We can expand the matrix $K^\pm$ on a basis of $SU(N_f)$ generators:

$$K^\pm = a_0 + a_i \tau_i$$

$$a_0 = \frac{1}{N_f} tr K^\pm = \frac{1}{N_f} \bar{\psi} \frac{1 \pm \gamma_5}{2} \psi = \frac{1}{2}(s_0 \mp i p_0)$$

$$a_a = \frac{1}{2} tr K^\pm \tau_a = \frac{1}{2} \bar{\psi} \tau_a \frac{1 \pm \gamma_5}{2} \psi = \frac{1}{2}(s_a \mp i p_a) \tag{3.20}$$

where we defined the following bilinear forms:

$$s_0 = \frac{2}{N_f} \bar{\psi}\psi \qquad p_0 = \frac{2}{N_f} \bar{\psi} i\gamma_5 \psi \qquad s_a = \bar{\psi}\tau_a\psi \qquad p_a = \bar{\psi} i\gamma_5 \tau_a \psi \tag{3.21}$$

Consider the case of two flavors: $N_f = 2$. The determinant of the 2×2 matrix (3.20) is given by (A.6):

$$\det K^\pm(x) = a_0^2 - a_a a_a = \frac{1}{4}\left((s_0 \mp i p_0)^2 - (s_a \mp i p_a)^2\right) \tag{3.22}$$

and the instanton induced interaction is:

$$I_{int}\left(\psi, \psi^\dagger\right)_{N_f=2} = -\frac{g_s^2}{2} \int d_4x \left[\left(\bar{\psi}\psi\right)^2 + \left(\bar{\psi} i\gamma_5 \vec{\tau} \psi\right)^2 - \left(\bar{\psi}\vec{\tau}\psi\right)^2 - \left(\bar{\psi} i\gamma_5 \psi\right)^2\right] \tag{3.23}$$

The coupling constant is :

$$g_s^2 = \frac{N}{V}\left(\frac{2M_0 V}{N}\right)^2 \tag{3.24}$$

with a dimension E^{-2} of an inverse energy squared. The interaction is attractive in the $\bar{\psi}\psi$ and $\bar{\psi} i\gamma_5 \vec{\tau} \psi$ channels, like the gluon exchange interaction (3.4) and the

4-quark interaction (4.4) of the Nambu Jona-Lasinio model. However, contrary to the latter, it is repulsive in the $\bar{\psi}\vec{\tau}\psi$ and $\bar{\psi}i\gamma_5\psi$ channels (see section 3.3).

In the case of three flavors, $N_f = 3$, the determinant of the matrix (3.20) is given by (A.6):

$$\det K^{\pm}(x) = \left(a_0^{\pm}\right)^3 - a_0^{\pm} a_a^{\pm} a_a^{\pm} + \frac{2}{3} d_{abc} a_a^{\pm} a_b^{\pm} a_c^{\pm} \qquad (N_f = 3) \tag{3.25}$$

Equation (3.22) shows that only even powers of p contribute to ($\det K^+ + \det K^-$). We obtain a six-quark interaction:

$$I_{int}\left(\psi, \psi^{\dagger}\right)_{N_f=3} = -\frac{N}{2V}\left(\frac{2M_0V}{N}\right)^3 \int d_4x \left\{\frac{1}{8}\left(s_0^3 - 3s_0p_0^2\right)\right.$$

$$\left. -\frac{1}{8}\left[s_0s_a^2 - s_0p_a^2 - 2p_0s_ap_a\right] + \frac{1}{12}d_{abc}\left(s_as_bs_c - s_ap_bp_c - p_as_bp_c - p_ap_bs_c\right)\right\} \tag{3.26}$$

3.3 Comparison between gluon exchange and instanton induced interactions

There are considerable differences between the effective gluon exchange interaction (3.4) and the instanton induced interactions (3.23) and (3.26). The instanton induced interaction has a non-locality which effectively regularizes the model. It does not generate vector interactions, at least not in the way it is presently derived [26, 27] . This suggests that, if indeed ρ and ω mesons are bound $q\bar{q}$ excitations, then the force which binds them is not an instanton induced interaction. Both interactions are attractive in the scalar and pseudoscalar channels $(\bar{q}q)$ and $(\bar{q}i\gamma_5\vec{\tau}q)$. This means they can both predict spontaneous chiral symmetry breaking and the appearance of pions as Goldstone bosons (see section 8.4). However, if we compare the interactions (3.4) and (3.23) we note that the gluon exchange interaction is attractive in the $(\bar{q}\vec{\tau}q)$ and $(\bar{q}i\gamma_5q)$ channels whereas the instanton induced interaction is repulsive in these channels. In these channels, bound $q\bar{q}$ states will not be formed. In this sense, the effective gluon exchange interaction (3.4) is closer to the interaction (4.4) of the Nambu Jona-Lasinio model.

In a recent and still controversial analysis of lattice QCD calculations in terms of cooling, instantons and anti-instantons appear as important contributions to the low- lying configurations of the euclidean action [34]. The instanton configurations seem to dominate the structure of the π, the nucleon and the ρ.

3.4 Non-local effective interactions

When certain degrees of freedom are integrated out, or eliminated using the equations of motion, the resulting action usually involves effective interactions which are non-local in both space and time. For example, attempts to eliminate, at least formally, the gluon degrees of freedom invariably lead to non-local effective interactions [14, 24], After bosonization, the effective theory is expressed in terms of colorless non-local fields which represent re-summations of gluon ladders. Non-local effective interactions have the advantage of eliminating ultraviolet divergences (see section 6.1).

A typical euclidean effective action, describing quarks interacting with non-local fields, can be written in the form:

$$I(S,q,\bar{q}) = \int d_4x\, \bar{q}(x)\left(-i\partial_\mu\gamma_\mu + m\right)q(x)$$

$$+\int d_4x d_4y\, \bar{q}(x)\left\langle x\left|\Gamma_a S_a\right|y\right\rangle q(y) + \frac{1}{2g^2}\int d_4x S_a^2(x) \tag{3.27}$$

This action describes quarks interacting with non-local fields $\langle x|S_a|y\rangle$. The operators $\Gamma_a = (1, i\gamma_5, \tau_a, i\gamma_5\tau_a, ...)$ distinguish the interactions of the quarks in various $q\bar{q}$ channels. The action is easily generalized to include vector interactions as well as different coupling constants.

Two forms have been considered for the non-local fields. The first form, used in [18, 35, 36] for example, is:

$$\langle x|S_a|y\rangle = S_a\left(\frac{x+y}{2}\right)r(x-y) \qquad \langle k|S_a|k'\rangle = \frac{1}{\beta\Omega}S_a(k'-k)\,r\left(\frac{k+k'}{2}\right) \tag{3.28}$$

In (3.28) the matrix elements of the non-local field are given both in the space–time and momentum representations, with $S_a(q) = \int d_4x e^{iq_\mu x_\mu} S_a(x)$ and with the plane waves defined in (3.31). When this form is used, the *dynamical variables* are considered to be the local fields $S_a(x)$ and not the regulator function $r(x)$. The latter expresses the non-locality of the interaction of the quarks with the fields $S_a(x)$ and, presumably, the degrees of freedom which have been integrated out. The dynamical variable $S_a(x)$ appears as a constraint in the action (3.27). It can be eliminated in favor of the quark degrees of freedom by writing its equation of motion:

$$\frac{1}{g^2}S_a(X) = -\int d_4x\, \bar{q}\left(X+\frac{x}{2}\right)\Gamma_a r(x)\, q\left(X-\frac{x}{2}\right) \tag{3.29}$$

We can substitute this expression back into the action (3.27) and write the resulting action in terms of quark fields alone. The result is often quoted in the literature in terms of quark fields $q(k)$ in the momentum representation:

$$q(x) = \sum_k \langle x|k\rangle\, q(k) \qquad \bar{q}(x) = \sum_k \langle k|x\rangle\, \bar{q}(k) \tag{3.30}$$

where $\langle x|k\rangle$ are the 4-dimensional plane waves:

$$\langle x \mid k\rangle = \frac{1}{\sqrt{\beta\Omega}} e^{ik_\mu x_\mu}, \quad \int d_4x \langle k \mid x\rangle \langle x \mid k'\rangle = \delta_{kk'}, \quad \sum_k \langle x \mid k\rangle \langle k \mid y\rangle = \delta(x-y) \tag{3.31}$$

with Ω being the volume of the system and β the inverse temperature. For a large volume and in the limit $\beta \to \infty$ of zero temperature, we have:

$$\frac{1}{\beta\Omega}\sum_k f(k) = \frac{1}{(2\pi)^4}\int d_4k f(k) \tag{3.32}$$

The action (3.27) becomes:

$$I(q,\bar{q}) = \int d_4x\, \bar{q}(x)(-i\partial_\mu\gamma_\mu + m)\, q(x)$$

$$-\frac{1}{2g^2}\frac{1}{\beta\Omega}\sum_{k_1k_2k_3k_4} \delta_{k_2-k_1,k_3-k_4}\left[\bar{q}(k_1)\Gamma_a r\left(\frac{k_1+k_2}{2}\right) q(k_2)\right]$$
$$\left[\bar{q}(k_3)\Gamma_a r\left(\frac{k_3+k_4}{2}\right) q(k_4)\right] \tag{3.33}$$

Another form for the non-local fields is obtained when interactions are expressed in terms of the delocalized quark fields (3.13) which appear, for example, in the instanton induced interactions. The interaction of the quarks with the non-local fields has the form:

$$\int d_4x\bar{\psi}(x)\Gamma_a S_a(x)\psi(x) = \int d_4x d_4y\; \bar{q}(x)\langle x|rS_a r|y\rangle\Gamma_a q(y) \tag{3.34}$$

where r is a regulator, which is diagonal in momentum space:

$$\langle k|r|k'\rangle = \delta_{kk'} r\left(k^2\right) \tag{3.35}$$

The regulator is a decreasing function of the euclidean k^2, as in (3.12) for example. Again, the local field $S_a(x)$ is assumed to be the dynamical variable, and not the regulator r. The form (3.34) is used in [26, 37, 38] for example. When we insert this form into the action (3.27), the field $S_a(x)$ may again be eliminated as above, and the action can be written in terms of quark variables thus:

$$I(q,\bar{q}) = \langle\bar{q}|(-i\partial_\mu\gamma_\mu + m)|q\rangle - \frac{1}{2g^2}\langle\bar{q}|r^2\Gamma_a|q\rangle^2 \tag{3.36}$$

where the quark fields are expressed in shorthand notation $q(x) = \langle x|q\rangle$ and $\bar{q}(x) = \langle\bar{q}|x\rangle$. If we express the action in terms of quark fields $q(k) = \langle k|q\rangle$ in momentum space, the action acquires the form (3.33) with $r\left(\frac{k_1+k_2}{2}\right)$ replaced by $r(k_1)r(k_2)$.The form (3.34) is more convenient to work with in the soliton sector because it leads to the genuine separable interaction displayed in (3.36). The form (3.28) is only separable in translationally invariant systems. In both cases, the non-locality regularizes the theory (see chapter 6).

4

THE NAMBU JONA-LASINIO MODEL

The Nambu Jona-Lasinio model aims at a unified description of the physical vacuum, of mesons which are $q\bar{q}$ excitations of the vacuum, and of baryons which are constructed as bound states of quarks, which are either solitons or quark–diquark substructures. The model makes use of an attractive interaction in the scalar $\bar{q}q$ channel, which is strong enough to cause spontaneous chiral symmetry breaking and which gives quarks a constituent mass. Like other models which account dynamically for the spontaneous breakdown of chiral symmetry, it is unable to confine color and quarks. Hadrons appear in the model as bound (instead of confined) states of quarks. The model was conceived before the discovery of quarks. It was originally designed to be a dynamical model of spontaneous chiral symmetry breaking in the vacuum [39, 40]. References to reviews and applications are given in chapter 10 for the meson sector and in chapters 12 and 13 for the baryon sector.

4.1 The lagrangian and the hamiltonian of the Nambu Jona-Lasinio model

In its simplest form, the Nambu Jona-Lasinio model describes a system of quarks with four-fermion interactions. The lagrangian is:

$$\mathcal{L} = \overline{q}\left(i\partial_\mu\gamma^\mu - m\right)q + \frac{g_s^2}{2}\left[\left(\overline{q}\tau_a q\right)^2 + \left(\overline{q}i\gamma_5\tau_a q\right)^2\right]$$

$$-\frac{g_v^2}{2}\left(\left(\overline{q}\gamma_\mu\tau_a q\right)\left(\overline{q}\gamma^\mu\tau_a q\right) + \left(\overline{q}\gamma_5\gamma_\mu\tau_a q\right)\left(\overline{q}\gamma_5\gamma^\mu\tau_a q\right)\right) \tag{4.1}$$

The four interactions will be referred to as the scalar, pseudoscalar, vector and axial interactions respectively. In section 4.5.3 an extra term will be added so as to account for the observed axial $U_A(1)$ anomaly. The dynamical quark fields are q and $\bar{q} = q^\dagger\beta$. The scalar and vector coupling constants, respectively g_s^2 and g_v^2, have the dimension E^{-2} of an inverse energy squared. Their sign is chosen such as to give an attraction in the $\sigma, \pi, \rho, \omega, ...$ quark–antiquark channels. The quark field q has dimension $E^{3/2}$. The four-fermion interactions are constructed out of the quark bilinear forms listed in Table 2.2. Further interactions could be added. The N_f^2 operators τ_a, for $a = 0, 1, ..., N_f^2$ act on the *flavor* variables of the quark fields. They are matrices of a $U(N_f)$ algebra (section A.2). When only u and d quarks are taken into account, they are the three Pauli matrices (A.21) together with the unit matrix $\tau_0 = I$. When u, d and s quarks are taken into account, they are the eight Gell-Mann matrices $\tau_1, \tau_2, ..., \tau_8$ (A.22) together with $\tau_0 = \sqrt{\frac{2}{3}} I$ where I is the unit matrix. The mass matrix $m = diag\,(m_u, m_d, m_s, ...)$ is diagonal in flavor space:

$$m = \sum_{i=u,d,s,...} |i\rangle\, m_i\, \langle i| \tag{4.2}$$

where m_i are the *current* quark masses. An estimate of current quark masses is given in Table 4.2 (section 4.5). The Minkowski 4-vectors in (4.1) are:

$$x^{\mu} = (t, \boldsymbol{r}) \qquad x_{\mu} = (t, -\boldsymbol{r}) \qquad \gamma^{\mu} = (\beta, \boldsymbol{\gamma}) \qquad \gamma_{\mu} = (\beta, -\boldsymbol{\gamma}) \tag{4.3}$$

By canonical quantization of the quark field in (4.1) we obtain the hermitian hamiltonian of the system:

$$H\left(q, q^{\dagger}\right) = \int d_3 r \left[q^{\dagger} \left(\frac{\boldsymbol{\alpha}.\boldsymbol{\nabla}}{i} + \beta m \right) q - \frac{g_s^2}{2} \int d_3 r \left(\left[\left(\overline{q} \tau_a q \right)^2 + \left(\overline{q} i \gamma_5 \tau_a q \right)^2 \right] \right) \right.$$

$$\left. + \frac{g_v^2}{2} \left[\left(\overline{q} \gamma_{\mu} \tau_a q \right) \left(\overline{q} \gamma^{\mu} \tau_a q \right) + \left(\overline{q} \gamma_5 \gamma_{\mu} \tau_a q \right) \left(\overline{q} \gamma_5 \gamma^{\mu} \tau_a q \right) \right] \right] \tag{4.4}$$

The Nambu Jona-Lasinio model leads to ultraviolet divergences. It is not defined until it is regularized in chapter 6.

4.2 The partition function and the euclidean action.

The partition function of the system, described by the hamiltonian (4.4), can be expressed in terms of the path integral [41, 42]:

$$Tr e^{-\beta H} \equiv \int D\left(q\right) D\left(q^{\dagger}\right) e^{-I\left(q, q^{\dagger}\right)} \tag{4.5}$$

where the action:

$$I\left(q, q^{\dagger}\right) = \int_0^{\beta} d\tau \left(q^{\dagger} \partial_{\tau} q + H\left(q, q^{\dagger}\right) \right) \tag{4.6}$$

is referred to as the *euclidean action* and where β is the inverse temperature. An introduction to path integrals over fermion fields is found in [41, 42] for example. The reason for calling the action 'euclidean' is explained in section 4.3. In the path integral (4.5), the quark fields q and $q^{\dagger}$ (which are independent dynamical variables) are independent anticommuting Grassmann variables, $H\left(q^{\dagger}, q\right)$ is the form (4.4) and β is the inverse temperature. At finite temperature, the quark fields need to satisfy antiperiodic boundary conditions (see section 10.5). The zero temperature limit is $\beta \to \infty$, in which case $e^{-\beta H} = |0\rangle\, e^{-\beta E_0}\, \langle 0|$ projects onto the ground state of the system.

Inserting the form (4.4) into the expression (4.5), we obtain an explicit expression of the partition function:

$$Tre^{-\beta H} \equiv \int D(q)\, D\left(q^{\dagger}\right) e^{-\int d_4 x q^{\dagger}\left(\partial_\tau + \frac{\alpha.\nabla}{i} + \beta m\right) q} e^{\frac{g_s^2}{2}\int d_4 x\left[(\bar{q}\tau_a q)^2 + (\bar{q} i\gamma_5 \tau_a q)^2\right]}$$

$$e^{-\frac{g_v^2}{2}\int d_4 x\left[(\bar{q}\gamma_\mu \tau_a q)(\bar{q}\gamma^\mu \tau_a q) + (\bar{q}\gamma_5\gamma_\mu \tau_a q)(\bar{q}\gamma_5 \gamma^\mu \tau_a q)\right]} \quad (4.7)$$

where:

$$\int d_4 x \equiv \int_0^\beta d\tau \int d_3 r \quad (4.8)$$

Euclidean actions are useful because it is almost always the partition function which we calculate in various approximation schemes. Perturbation theory, expressed in terms of Feynman graphs, is derived from the path integral in chapter 8.

We repeatedly make use of the gaussian integrals:

$$\int \prod_i dz_i dz_i' e^{-z_i' A_{ij} z_j + u_i z_i + v_i z_i'} = (\det A)\, e^{u_i A_{ij}^{-1} v_j} = e^{Tr \ln A + u_i A_{ij}^{-1} v_j} \quad (4.9)$$

$$\int_{-\infty}^{\infty} \prod_i \left(\frac{dx_i}{\sqrt{2\pi}}\right) e^{-\frac{1}{2}x_i A_{ij} x_j + b_i x_i} = \frac{1}{\sqrt{\det A}} e^{\frac{1}{2} b_i A_{ij}^{-1} b_j} = e^{-\frac{1}{2} Tr \ln A + \frac{1}{2} b_i A_{ij}^{-1} b_j} \quad (4.10)$$

In the equation (4.9), z, z', u and v are Grassmann variables and A is an antisymmetric matrix. In the equation (4.10), x and b are real variables and A is a symmetric matrix. More general gaussian integrals are listed in Ref.[42].

4.3 Bosonization of the euclidean action

Let us write the hamiltonian $H\left(q, q^{\dagger}\right)$ in the form:

$$H\left(q, q^{\dagger}\right) = \int d_3 r\, q^{\dagger}\left(\frac{\alpha.\nabla}{i} + \beta m\right) q + V \quad (4.11)$$

where V contains the interactions, which are quartic in the quark fields:

$$V = -\int d_3 r \left[\frac{g_s^2}{2}\left((\bar{q}\tau_a q)^2 + (\bar{q} i\gamma_5 \tau_a q)^2\right) + \frac{g_v^2}{2}\left((\bar{q}\gamma\tau_a q)^2 + (\bar{q}\gamma_5 \gamma\tau_a q)^2\right)\right]$$

$$+ \int d_3 r\, \frac{g_v^2}{2}\left(\left(q^{\dagger}\tau_a q\right)^2 + \left(q^{\dagger}\gamma_5 \tau_a q\right)^2\right) \quad (4.12)$$

We have purposely separated out the time components $\left(\bar{q}\gamma^0 \tau_a q\right) = \left(q^{\dagger}\tau_a q\right)$ of the vector interactions, because they occur with an opposite sign.

The path integral formulation (4.5) of the partition function involves an integral which is quartic in the quark fields and which we cannot perform. The bosonization procedure consists in introducing extra fields, so as to make the exponential quadratic in the quark fields. The latter can then be integrated out using gaussian integration. The adequate identities are given in appendix C. For the first term of (4.12), we can use the identity (C.2) to write:

$$e^{\frac{g_s^2}{2}\int d_4x(\bar{q}\tau_a q)^2} = \frac{\int D(S)\, e^{-\int d_4x\left(\frac{1}{2g_s^2}S_a^2 + S_a(\bar{q}\tau_a q)\right)}}{\int D(S)\, e^{-\int d_4x\left(\frac{1}{2g_s^2}S_a^2\right)}} \tag{4.13}$$

This is a linearization of the four-quark interaction in the sense that, at the expense of introducing the extra fields S_a, the right hand side exponential is now quadratic in the quark fields, instead of being quartic. We associate in this way real auxiliary fields to the quark bilinears appearing in the four first terms of (4.12). They are the fields S_a, P_a, V_{ia} and A_{ia} indicated in Table 4.1 (section 4.4).

The last two terms of (4.12) have an opposite sign. We use the identity (C.4) to linearize them with a pure imaginary auxiliary field. For example:

$$e^{-\int d_4x\left(\frac{g_v^2}{2}\left(q^\dagger\tau_a q\right)^\dagger\left(q^\dagger\tau_a q\right)\right)} = \frac{\int D(V_4)\, e^{-\frac{1}{2g_v^2}V_{4a}^2 - iV_{4a}\left(q^\dagger\tau_a q\right)}}{\int D(V_4)\, e^{-\frac{1}{2g_v^2}V_{4a}^2}} \tag{4.14}$$

We thus associate imaginary fields iV_{4a} and iA_{4a} to the bilincars $\left(q^\dagger\tau_a q\right)$ and $\left(q^\dagger\gamma_5\tau_a q\right)$.

We then substitute the expressions (4.13) and (4.14) into the partition function (4.7), which acquires the form:

$$Tre^{-\beta H} \equiv \frac{1}{N}\int D(q)\, D\left(q^\dagger\right) D(S)\, D(P)\, D(V)\, D(A)\, e^{-I\left(q,q^\dagger,S,P,V,A\right)} \tag{4.15}$$

where the action is:

$$I\left(q,q^\dagger,S,P,V,A\right) = \int d_4x q^\dagger\left(\partial_\tau + h\right) q$$

$$+\frac{1}{2g_s^2}\int d_4x\left(S_a^2 + P_a^2\right) + \frac{1}{2g_v^2}\int d_4x\left(\boldsymbol{V}_a^2 + V_{4a}^2 + \boldsymbol{A}_a^2 + A_{4a}^2\right) \tag{4.16}$$

and where h is the Dirac hamiltonian:

$$h = \frac{\boldsymbol{\alpha}.\boldsymbol{\nabla}}{i} + \beta m + \beta\left(S_a + i\gamma_5 P_a\right)\tau_a + \boldsymbol{\alpha}.\boldsymbol{V}_a\tau_a + iV_{4a}\tau_a - \gamma_5\boldsymbol{\alpha}.\boldsymbol{A}_a\tau_a - i\gamma_5 A_{4a}\tau_a \tag{4.17}$$

which is non-hermitian because of the time components V_{4a} and A_{4a} of the vector fields. In the meson sector, the use of imaginary time components of the vector fields is a very useful way to separate the action into real and imaginary parts. The latter gives rise to the so-called anomalous processes, which are known to fit experiment when they are regularized with an infinite cut-off and which is often left unregularized, as discussed in section 6.3. It causes problems however in the soliton sector, as discussed in section 12.10. The constant N is:

$$N = \int D(S)\,D(P)\,D(V)\,D(A)\,e^{-\left(\frac{1}{2g_s^2}\int d_4x\left(S_a^2+P_a^2\right)+\frac{1}{2g_v^2}\int d_4x\left(\boldsymbol{V}_a^2+V_{4a}^2+\boldsymbol{A}_a^2+A_{4a}^2\right)\right)} \tag{4.18}$$

and the auxiliary fields $S, P, \boldsymbol{V}, \boldsymbol{A}, V_4$ and A_4 are real integration variables. In the semi-bosonized action (4.16), the quark fields appear in quadratic form. In the partition function (4.15), the integral over the Grassmann variables q and $q^\dagger$ is therefore gaussian and the quark fields may be integrated out using (4.9). The partition function acquires then the form:

$$Tre^{-\beta H} \equiv \frac{1}{N}\int D(S)\,D(P)\,D(V)\,D(A)\,e^{-I(S,P,V,A)} \tag{4.19}$$

where $I(S,P,V,A)$ is the so-called *bosonized euclidean* action:

$$I(S,P,V,A) = -Tr\ln(\partial_\tau + h)$$

$$+\frac{1}{2g_s^2}\int d_4x\left(S_a^2+P_a^2\right)+\frac{1}{2g_v^2}\int d_4x\left(\boldsymbol{V}_a^2+V_{4a}^2+\boldsymbol{A}_a^2+A_{4a}^2\right) \tag{4.20}$$

The trace in (4.20) is taken over the space spanned by the eigenvalues of the operator $\partial_\tau + h$, or equivalently, over the variables of the quark field q. The term $-Tr\ln(\partial_\tau + h)$ is often referred to as the *quark loop term* or as the *fermion determinant.* The quark loop term represents a system of quarks, interacting with the fields S, P, V and A.

The field $S+i\gamma_5 P$ is often referred to as a *chiral field.* It is often expressed in the form $S+i\gamma_5 P = \varphi e^{i\gamma_5\theta}$ where $\varphi \equiv \varphi_a\tau_a$ and $\theta \equiv \theta_a\tau_a$ are hermitian. In a chiral transformation as defined by (4.30) (section 4.5.2), $\frac{1}{2}tr_f\,(S+i\gamma_5 P)^\dagger(S+i\gamma_5 P)$ $= S_a^2 + P_a^2$ remains invariant. In chiral transformation, the fields S and P are said to remain on the *chiral circle*, defined by the radius $S_a^2 + P_a^2$.

It is often more convenient to use a covariant *form* of the action (4.20). It is obtained by defining the following 4-vectors *with a euclidean metric*:

$$x_\mu = x^\mu = (x_4, x_i) = (\tau, \boldsymbol{r}) \qquad \gamma_\mu = \gamma^\mu = (\gamma_4, \boldsymbol{\gamma}) = (i\beta, \boldsymbol{\gamma}) = -(\gamma_\mu)^\dagger$$

$$V_{\mu a} = V^{\mu a} = (V_{4a}, \boldsymbol{V}_a) \qquad A_{\mu a} = A^{\mu a} = (A_{4a}, \boldsymbol{A}_a) \tag{4.21}$$

It is the euclidean metric $g^{\mu\nu} = g_{\mu\nu} = \delta_{\mu\nu}$ of the 4-vectors (4.21) which is the origin of the term 'euclidean action'. We can define a Dirac operator $D \equiv$

$\beta(\partial_\tau + h)$, where h is the Dirac hamiltonian (4.17). It can be written in terms of the euclidean 4-vectors (4.21) as follows:

$$D \equiv \beta(\partial_\tau + h) = -i\partial_\mu\gamma_\mu + m + S_a\tau_a + i\gamma_5 P_a\tau_a + \gamma_\mu V_{\mu a}\tau_a + \gamma_5\gamma_\mu A_{\mu a}\tau_a \tag{4.22}$$

The action (4.20) thus acquires the 'covariant' form:

$$I(S,P,V,A) = -Tr\ln\left(-i\partial_\mu\gamma_\mu + m + S_a\tau_a + i\gamma_5 P_a\tau_a + \gamma_\mu V_{\mu a}\tau_a + \gamma_5\gamma_\mu A_{\mu a}\tau_a\right)$$

$$+\frac{1}{2g_s^2}\int d_4x\left(S_a^2 + P_a^2\right) + \frac{1}{2g_v^2}\int d_4x\left(V_{\mu a}^2 + A_{\mu a}^2\right) \tag{4.23}$$

In writing (4.23), we have omitted the constant $-Tr\ln\beta$, which does not modify the dynamics of the system.

The form (4.23) is convenient for two reasons. First, because use can be made of the relations:

$$tr\gamma_\mu\gamma_\nu = tr\gamma^\mu\gamma^\nu = -4\delta_{\mu\nu} \qquad tr\gamma_\mu\gamma_\nu\gamma_\alpha\gamma_\beta = 4\left(\delta_{\mu\nu}\delta_{\alpha\beta} - \delta_{\mu\alpha}\delta_{\nu\beta} + \delta_{\mu\beta}\delta_{\nu\alpha}\right)$$

$$\gamma_5 = \gamma_1\gamma_2\gamma_3\gamma_4 \qquad Tr\gamma_5\gamma_\mu\gamma_\nu\gamma_\alpha\gamma_\beta = 4\epsilon_{\mu\nu\alpha\beta} \quad \left(\epsilon^{1234} = \epsilon_{1234} = 1\right) \tag{4.24}$$

which are useful when we expand the quark loop term in powers of the fields. Note that both the antisymmetric tensors ε^{1234} and ε_{1234} are positive. Second, only the imaginary part of the action contributes to anomalous processes (see section 6.3).

Note that the euclidean actions (4.16) or (4.23) have *not* been derived from a Lorentz invariant action by the substitutions $\tau \to it$, $V_4 \to iV_0$, and $e^{-\tau H} \to e^{-iHt}$. Euclidean actions are obtained directly by deriving path integral representations of the partition function. Lorentz invariant actions, involving 4-vectors in a Minkowski space, occur in path integral representations of matrix elements of the evolution operator e^{-iHt}. Path integral representations of the partition function $Tre^{-\beta H}$ need not be derived by some analytic continuation of path integrals representing matrix elements of the evolution operator e^{-iHt}, as is often claimed.

The auxiliary fields acquire a physical meaning when the path integral (4.19) is evaluated by the stationary phase method or the method of steepest descent. At a stationary point of the action, the auxiliary fields become the mean fields of the Hartree approximation (section 7.4.3). Thus:

$$S_a(x) = -g_s^2\left\langle\bar{q}(x)\tau_a q(x)\right\rangle, \ldots, \quad iV_{4a}(x) = g_v^2\left\langle q^\dagger(x)\tau_a q(x)\right\rangle, \ldots \tag{4.25}$$

where $\langle\rangle$ denotes the expectation value in the ground state, which can be either the vacuum or a soliton. It is not surprising that the stationary point occurs for imaginary values of the time components of the vector fields, since they represent expectation values of hermitian operators (see the discussion in section 12.10).

Table 4.1 *Constraints relating real fields to hermitian bilinear operators*

$S_a(x) = -g_s^2 \bar{q}(x) \tau_a q(x)$	$P_a(x) = -g_s^2 \bar{q}(x) i\gamma_5 \tau_a q(x)$
$\vec{V}_a(x) = -g_v^2 \bar{q}(x) \vec{\gamma} \tau_a q(x)$	$iV_{4a}(x) = g_v^2 q^\dagger(x) \tau_a q(x)$
$\vec{A}_a(x) = -g_v^2 \bar{q}(x) \gamma_5 \vec{\gamma} \tau_a q(x)$	$iA_{4a}(x) = g_v^2 q^\dagger(x) \gamma_5 \tau_a q(x)$

4.4 Three equivalent descriptions of the system

In the preceding section, we have described the system in three equivalent ways. The partition function (4.7) and the action $I\left(q, q^\dagger\right)$, defined in (4.6), describe the system in terms of quark degrees of freedom. The partition function (4.15) and the semi-bosonized action $I\left(q, q^\dagger, S, P, V, A\right)$, defined in (4.16), describe the system in terms of both the quark degrees of freedom and the auxiliary fields S, P, V, A. This intermediate description involves redundant fields. Indeed, in the action (4.16), no time derivatives of the fields S, P, V, A appear, so that the latter are constraints rather than dynamical variables. The constraints can be expressed by making the action stationary with respect to the fields. They relate each real field to one of the hermitian bilinear operators, Table 4.1.

However, once the quark variables are integrated out, as in (4.19), the system is described by the bosonized action $I(S, P, V, A)$, defined in (4.20). In this formulation, the fields S, P, V, A become the dynamical variables of the system. Indeed, time derivatives of the fields S, P, V, A appear at all orders in the action (4.20). They are generated by the term $-Tr \ln(\partial_\tau + h)$. The occurrence of time derivatives of order higher than two in the effective action $I(S, P, V, A)$ reflects the fact that the fields S, P, V, A represent *composite* $q\bar{q}$ systems displayed in Table 4.1. The number of degrees of freedom expressed by the fields S, P, V, A is less than the number of degrees of freedom represented by the $q\bar{q}$ systems. This is why Skyrme type lagrangians, which involve only field variables, such as S, P, V, A, and which in addition are limited to second order derivatives of the fields, lack the proper number of degrees of freedom required to describe the quark structure of hadrons. They must always be supplemented by further terms involving higher order derivatives of the fields. The Wess–Zumino term is a typical example of a term which is added to describe, for example, the proper baryon number and density of the system.

It is not clear which of the actions (if any) would be obtained, were we able to derive a low energy effective action from QCD. In some derivations, which appeal to the instanton structure of the vacuum [26], the effective action is expressed in terms of the quark variables, as in (3.23), for example. In other derivations [18, 19], an effective action of the form (4.16) is derived. The form (4.16) is a convenient starting point because a purely quark theory can be derived from it by integrating out the fields S, P, V and A, while a pure mesonic representation of the system can be obtained by integrating out the quark fields q and $q^\dagger$.

4.5 Symmetries of the Nambu Jona-Lasinio model

Models of hadrons should implement the symmetries of strong interactions. Lorentz invariance, parity, time reversal, charge conjugation, G-parity, flavor and chiral symmetry are the symmetries of strong interactions which were known before the advent of QCD and which, of course, are also symmetries of QCD. However, QCD has an additional color gauge symmetry. The Nambu Jona-Lasinio model lacks color gauge symmetry, although the action is invariant under global color symmetry.

The Nambu Jona-Lasinio action (4.1) is invariant under Lorentz transformations, parity, time reversal, charge conjugation and G parity, which are symmetries of strong interactions. This can be checked by standard methods (see [5] for example).

Since the hamiltonian is the time component of the Lorentz four-vector $(H, \vec{p})$, the partition function $Tre^{-\beta H}$ is not invariant under Lorentz transformations. The 'covariant' form (4.23) of the euclidean action is, of course, also not invariant under Lorentz transformations. However, we shall see that meson correlation functions calculated from the euclidean action (4.23) do lead to Lorentz invariant on-shell meson masses.

Flavor and chiral symmetry are approximate symmetries of strong interactions. In QCD they are broken by the current quark masses (and also by electroweak interactions). The current quark masses, summarized in Table 4.2, are quoted from the 1996 Particle Data, [8]. Since quarks are confined, they do not propagate freely and the experimental determination of their masses is model-dependent. The determination of the light u and d current quark masses requires an independent determination of the quark condensates. This is usually provided by a somewhat model-dependent estimate from QCD sum-rules [43, 44, 45]. A recent discussion and compilation of current quark masses and quark condensates derived from lattice calculations is given in [46]. In the Standard Model the current quark masses are due to the interaction of the quarks with the Higgs field [47]. We do not even know whether this interaction can be accounted for in terms of the strong, electroweak and gravitational forces, or whether it is a force of a different kind. In QCD, invariance with respect to flavor rotations follows from the fact that the quark–gluon interaction is flavor-independent. Gluon exchange is the same for quarks of any flavor. Invariance with respect to flavor rotations is broken by the current quark masses and by electromagnetic interactions.

By introducing the mass matrix (4.2) into the Nambu Jona-Lasinio action, we have tacitly assumed that current quark mass terms are also responsible for the effect of broken flavor and chiral symmetry in the low energy theory. This assumption has not been justified.

In the chiral limit $m \to 0$ of vanishing current quark masses, the Nambu Jona-Lasinio hamiltonian (4.4) and the euclidean action (4.6) are invariant under flavor rotations and chiral transformations, defined in section 2.2. This invariance is maintained when auxiliary fields are introduced into the bosonized action, provided that the latter have definite transformation properties. The transfor-

Table 4.2 *Current quark masses [8]*

Quark	Flavor	Mass	Charge
Down	d	$5 - 15$ MeV	$-\frac{e}{3}$
Up	u	$2 - 8$ MeV	$\frac{2e}{3}$
Strange	s	$100 - 300$ MeV	$-\frac{e}{3}$
Charmed	c	$1 - 1.6$ GeV	$\frac{2e}{3}$
Bottom	b	$4.1 - 4.5$ GeV	$-\frac{e}{3}$
Top	t	180 ± 12 GeV	$\frac{2e}{3}$

mation properties of the action (4.23) are most easily expressed in terms of the fields:

$$S = S_a \tau_a \qquad P = P_a \tau_a \qquad V = \gamma_\mu V_{\mu a} \tau_a \qquad A = \gamma_\mu A_{\mu a} \tau_a \tag{4.26}$$

The action (4.23) can be written in the form:

$$I(S, P, V, A) = -Tr \ln\left(-i\partial_\mu \gamma_\mu + m + S + i\gamma_5 P + V + \gamma_5 A\right)$$

$$+\frac{1}{4\nu g_s^2} Tr\,(S + i\gamma_5 P)(S - i\gamma_5 P) + \frac{1}{4\nu g_v^2} Tr\left(VV^\dagger + AA^\dagger\right) \tag{4.27}$$

where:

$$\nu = 2\,(spin) \times N_c\,(color) \times N_f\,(flavor) \tag{4.28}$$

is the spin, color and flavor degeneracy of the quarks.

4.5.1 *Flavor symmetry*

The difference between the masses of the u and d quarks is small (a few MeV) relative to the masses (several hundred MeV) of hadrons formed by these quarks. In the limit $m_u = m_d$ where these masses are set equal, the action (4.27) becomes invariant with respect to *isospin rotations*, defined in section 2.2. An isospin rotation can be performed by the unitary operator $U = \exp\left(-\frac{i}{2}\sum_{a=1}^{3} \alpha_a \tau_a\right)$ acting on the flavor variables of the u and d quarks only. An isospin rotation is defined by the following transformations:

$$q \to Uq \qquad q^\dagger \to q^\dagger U^\dagger \quad \left(\bar{q} \to \bar{q} U^\dagger\right)$$

$$S + i\gamma_5 P + V + A \to U\,(S + i\gamma_5 P + V + A)\,U^\dagger \tag{4.29}$$

Invariance with respect to isospin rotations gives rise to near degenerate isospin multiplets. Isospin rotations form a $SU(2)_f$ group. The degeneracy of the $SU(2)_f$

multiplets is lifted by a few MeV due to the small difference between the u and d masses and mainly due to electromagnetic interactions [48].

In the limit $m_u = m_d = m_s$ where the masses of the u, d and s quarks are set equal, the action (4.27) becomes invariant under *flavor rotations.* A flavor rotation is defined by the transformations (4.29) in which the unitary operator has the form $U = \exp\left(-\frac{i}{2}\sum_{a=1}^{8}\alpha_a\tau_a\right)$ and acts on the flavor variables of the u, d and s quarks. Flavor rotations form an $SU(3)_f$ group. The current mass of the strange quark is about 200 MeV higher than the current masses of the u and d quarks. This is why the mass splittings (several hundred MeV) between the $SU(3)_f$ multiplets are higher than the mass splittings (a few MeV) of the isospin $SU(2)_f$ multiplets. In QCD, invariance with respect to flavor rotations follows from the fact that the quark–gluon interaction is flavor-independent. Gluon exchange is the same for quarks of any flavor. Invariance with respect to flavor rotations is broken by the flavor dependence of the current quark masses.

4.5.2 *Chiral symmetry*

This is not true of *chiral symmetry,* which is not only broken by the current quark masses but also dynamically by the strong interactions. A chiral transformation (section 2.2) can be performed by the unitary operator $U_A = \exp\left(-\frac{i}{2}\gamma_5\sum_{a=1}^{3}\alpha_a\tau_a\right)$ which acts on the flavor variables of the u and d quarks only. A chiral transformation is defined by the following transformations:

$$q \to U_A q \qquad q^\dagger \to q^\dagger U_A^\dagger \qquad \bar{q} \to \bar{q} U_A$$

$$S + i\gamma_5 P + V + A \to U_A^\dagger \left(S + i\gamma_5 P + V + A\right) U_A^\dagger \tag{4.30}$$

In the limit $m_u = m_d = 0$, a global chiral transformation (4.30) leaves the action (4.27) invariant. The limit $m_u, m_d \to 0$ is called the *chiral limit.* We shall see that the vacuum state is not invariant with respect to chiral transformations even in the chiral limit. We say that chiral symmetry is *spontaneously broken* in the vacuum. This is precisely what the Nambu Jona-Lasinio model was originally designed to do [39, 40]. The light mass (138 MeV) of the pion is a manifestation of the spontaneously broken chiral symmetry. In the chiral limit $m \to 0$, the pion becomes a massless particle, called a *Goldstone boson* (see chapter 8). A weaker chiral symmetry is often considered, which is invariance with respect to chiral transformations generated by the operator $U_A = \exp\left(-\frac{i}{2}\gamma_5\sum_{a=1}^{8}\alpha_a\tau_a\right)$ which acts on the flavor variables of the u, d and s quarks. Invariance of the action (4.27) with respect to such chiral transformations occurs in the limit of vanishing u, d and s current quark masses. In this limit, all the members of the $SU(3)_f$ octet are Goldstone bosons. Flavor and chiral transformations can also be expressed in terms of right and left transformations, discussed in section 2.3.

Note that the form (4.1) of the lagrangian is not entirely dictated by chiral symmetry. Chiral symmetry would not be broken if we chose independent coupling constants for the isoscalar vector and isoscalar axial interactions:

$$-\frac{g_{vs}^2}{2}(\bar{q}\gamma_\mu q)(\bar{q}\gamma^\mu q) - \frac{g_{va}^2}{2}(\bar{q}\gamma_5\gamma_\mu q)(\bar{q}\gamma_5\gamma^\mu q) \tag{4.31}$$

The near degenerate masses $\rho(770), \omega(782)$ and $a_1(1260), f_1(1285)$, shown in Table 2.2, are however obtained if one chooses equal coupling constants $g_v^2 = g_{vs}^2 = g_{va}^2$, as assumed in the lagrangian (4.1).

4.5.3 $U_A(1)$ *symmetry*

Table 2.1 shows that, in the chiral limit $m \to 0$ of vanishing quark masses, the Nambu Jona-Lasinio lagrangian (4.1) is invariant under axial $U_A(1)$ transformations (2.4). In the chirally symmetric broken phase, it would give rise to N_f^2 Goldstone bosons. For $N_f = 3$, they would constitute the nonet of π, K, η and η' mesons listed in Table 2.2. However, the η' has a mass of 958 MeV which makes it an unlikely Goldstone boson. This suggests that the physical vacuum has a broken $U_A(1)$ symmetry, even in the chiral limit. The breaking of $U_A(1)$ symmetry in QCD has been expressed in terms of a $U_A(1)$ anomaly, which can be related to the formation of instantons [32, 49, 50]. This yields anomalous contributions to the η and η' masses. The $U_A(1)$ symmetry is explicitly broken by instanton induced interactions which have the form of a 't Hooft determinant (see section 3.2). However, in order to account for the broken $U_A(1)$ symmetry in the Nambu Jona-Lasinio model (4.1), we need to add a phenomenological term which can take the form of a 't Hooft determinant:

$$V_{\text{det}} = \kappa \int d_4x \left(\det K^+ + \det K^-\right) \tag{4.32}$$

where K^+ and K^- are $N_f \times N_f$ matrices:

$$K_{ij}^{\pm} = \bar{q}_j \frac{1 \pm \gamma_5}{2} q_i \tag{4.33}$$

where i and j are flavor indices.

Under an axial $U_A(1)$ transformation, the matrices $K^{\pm}$transform as:

$$K^{\pm} \to e^{-i\theta} K^{\pm} e^{-i\theta} \tag{4.34}$$

As a result, the interaction V_{det} breaks $U_A(1)$ symmetry. For $N_f = 3$, the interaction (4.32) is a local six-quark interaction of the *form* (3.26). The form (4.32) is used in [29, 51, 52]. Instead of breaking $U_A(1)$ symmetry by adding the quark interaction (4.32), we can add to the bosonized or semi-bosonized actions (4.16) and (4.23) a $U_A(1)$ symmetry breaking term of the form:

$$V_{U1} = -\frac{\kappa^2}{2}\left(\det(S + iP) + \det(S - iP)\right) \tag{4.35}$$

where S and P are the $N_f \times N_f$ matrices defined in (4.26). In an axial transformation (2.4), the matrices S and P transform as follows:

$$S + iP \to e^{-i\theta} (S + iP) e^{-i\theta} \qquad S - iP \to e^{i\theta} (S - iP) e^{i\theta} \tag{4.36}$$

so that the interaction V_{U1} is not invariant. Acting in the meson sector, the interaction (4.35) produces flavor mixing, the effects of which have been estimated in [51]. A simpler $U_A(1)$ symmetry breaking interaction has been suggested by Schechter [53]:

$$V_{ven} = -\frac{\kappa^2}{2} \left(tr_f \ln (S + iP) - tr_f \ln (S - iP) \right)^2 \tag{4.37}$$

where tr_f is a trace over the flavor indices. The term (4.37) does not break chiral symmetry. In the meson sector, it does not produce flavor mixing. It simply adds extra mass terms for the $\eta(547)$ and $\eta'(958)$ (see section 5.6.1). Other forms which do not produce flavor mixing have been used [54, 55].

4.6 The case of large quark masses

From Table 4.2, we see that the current quark masses of the c, b and t quarks are respectively 1.35, 5 and 175 GeV. As a result, invariance with respect to flavor rotations is so strongly broken that the expression of the quark interactions in terms of $SU(N_f)$ generators is no longer useful. Furthermore, a new symmetry arises which effectively decouples the spin of the heavy quarks. We examine them in turn. The lagrangians written in this section are written in terms of Lorenz 4-vectors with a Minkowski metric. They are *not* euclidean.

4.6.1 *Suppression of the $SU(N_f)$ generators*

So far, we have written the Nambu Jona-Lasinio lagrangian in terms of the generators τ_a of the $SU(N_f)$ group. However, when the current quark masses are large, the flavor and chiral symmetry is strongly broken by the current quark mass term $m = diag(m_u, m_d, m_s, m_c, m_b, m_t)$ and the corresponding rotations are no longer interesting. We intuitively expect, for example, that the use of the 15 generators τ_a of the $SU(4)_f$ group is not very useful when a 1.35 GeV charmed quark is added to the u, d and s quarks. One might also want to consider the strange quark s to be 'heavy' so as to consider only $SU(2)$ flavor and chiral transformations of the u and d quarks. In such cases, it is useful to express the Nambu Jona-Lasinio lagrangian (4.1) directly in terms of the $u, d, s, ...$ quark fields, without having recourse to the $SU(N_f)$ matrices τ_a. This does not, of course, modify the symmetries of the action. The flavor generators τ_a can be eliminated by using the completeness relation (A.20):

$$\delta_{il}\delta_{kj} = \frac{1}{2} (\tau_a)_{ij} (\tau_a)_{kl} \tag{4.38}$$

in which the sum over the index a includes the unit matrix $\tau_{a=0} = \sqrt{\frac{2}{N_f}} I$ and in which $i, j, ...$ are flavor indices.. We write the Nambu Jona-Lasinio lagrangian (4.1) in the form:

$$\mathcal{L} = \overline{q}\left(i\partial_\mu \gamma^\mu - m\right) q - \mathcal{L}_{int} \tag{4.39}$$

where:

$$\mathcal{L}_{int} = -\frac{g_s^2}{2}\left[\left(\overline{q}\tau_a q\right)^2 + \left(\overline{q}i\gamma_5\tau_a q\right)^2\right] \tag{4.40}$$

Vector interactions can easily be added. We can use (4.38) to write the interaction $\mathcal{L}_{int}$ in the form:

$$\mathcal{L}_{int} = -g_s^2 \sum_{i,j=u,d,s,\dots} \left[\left(\overline{q}_i q_j\right)\left(\overline{q}_j q_i\right) + \left(\overline{q}_i i\gamma_5 q_j\right)\left(\overline{q}_j i\gamma_5 q_i\right)\right] \tag{4.41}$$

By making use of the completeness relation of the three Pauli matrices of the $SU(2)_f$ subgroup, for example, it would be quite simple to rewrite (4.41) so as to maintain explicit isospin invariance.

4.6.2 *Heavy quark spin symmetry*

The high masses of the c, b and t quarks leads to several simplifications in the dynamics of systems involving heavy quarks (see [56, 57, 58] and references therein). In this section we discuss some consequences for the heavy quark spectroscopy [59].Consider, for example, mesons composed of a heavy c or b quark and a light u or d antiquark. The $D(1870)$ and $D^*(2010)$ are bound $c\bar{u}$ and $c\bar{d}$ excitations with quantum numbers $L=0, S=0$ and $L=0, S=1$ respectively. The $B(5280)$ and $B^*(5330)$ are $u\bar{b}$ and $d\bar{b}$ excitations with the same quantum numbers. The small mass difference between the $S=1$ and $S=0$ mesons suggests that the spin of the heavy quark decouples from the dynamics as its mass increases. Consider the effective gluon exchange interaction (3.3) between c and u quarks:

$$\mathcal{L}_{int}(x) = -\frac{\alpha^2}{\Lambda^2}\left[\bar{c}\gamma^\mu\lambda_c c\right]\left[\bar{u}\gamma_\mu\lambda_c u\right] \tag{4.42}$$

Since the c quark has a mass as high as 1.5 GeV, it is useful to distinguish the upper and lower components of the heavy quark field:

$$h = P_+ c \qquad l = P_- c \qquad P_\pm = \frac{1}{2}\left(1 \pm v_\mu\gamma^\mu\right) \tag{4.43}$$

where v^μ is a given velocity vector with norm $v_\mu v^\mu = 1$. The operators $P_\pm$ are orthogonal projectors. For an on-shell quark with momentum p^μ we could take $v^\mu = p^\mu/m$. We consider a c quark at rest and we take $v^\mu = (1,0,0,0)$. In this case $P_\pm = \frac{1}{2}(1 \pm \beta)$ and the operators $P_\pm$ project on the upper and lower components of the Dirac field. Because of the large mass gap of the heavy quark, the lower component $l = P_- c$ of the heavy quark field can be neglected. Using the identities $P_+\beta P_+ = \beta$ and $P_+\boldsymbol{\gamma}P_+ = \boldsymbol{\gamma}$, we can reduce the interaction (4.42) to the form:

$$\mathcal{L}_{int}(x) = -\frac{\alpha^2}{\Lambda^2}\left[c^\dagger\lambda_c c\right]\left[u^\dagger\lambda_c u\right] \tag{4.44}$$

The interaction (4.44) is invariant in a rotation of the quark spin.

$$c \to e^{-i\sigma_i\theta_i} c \qquad\qquad c^\dagger \to c^\dagger e^{i\sigma_i\theta_i} \tag{4.45}$$

The heavy quark lagrangian can similarly be reduced to:

$$\mathcal{L}_c = \bar{c}\,(i\partial_\mu\gamma^\mu - m_c)\,c = c^\dagger\,(i\partial_t - m_c)\,c \tag{4.46}$$

which is invariant in the rotation (4.45) of the heavy quark spin. (The mass term can be eliminated by the substitution $c \to e^{-im_c t}c$.) It follows that a rotation of the heavy quark spin is asymmetry of the action and of the hamiltonian. Spectroscopic calculations of D and B mesons in this framework can be found in [59].

5

THE EQUIVALENT LINEAR SIGMA MODEL

The aim of this chapter is to transform the Nambu Jona-Lasinio action into the action of an equivalent linear sigma model. The equivalence holds when the cut-off which regulates the quark loop is large compared to the constituent quark masses. We shall find in fact that the cut–off is only about twice the constituent quark mass. In spite of this, the equivalent sigma model is a convenient way to understand the physics contained in the model and to make simple and realistic estimates of the model parameters. However, the equivalent sigma model hides many of the shortcomings of the model such as, for example, its inability to confine the quarks. The real physical content as well as the limitations of the Nambu Jona-Lasinio model become apparent when correlation functions and solitons are calculated (see chapters 7 and 9). This chapter also serves as a pedagogical introduction to some of the basic techniques used in the meson sector.

5.1 The Nambu Jona-Lasinio action with scalar fields

We start from the bosonized euclidean action (4.23) and we neglect the vector interactions by setting $V = A = 0$:

$$I(S,P) = -Tr\ln\left(-i\partial_\mu\gamma_\mu + m + S + i\gamma_5 P\right) + \frac{1}{2g_s^2}\int d_4x\left(S_a^2 + P_a^2\right) \tag{5.1}$$

where m is the current quark mass matrix (4.2). A $U_A(1)$ symmetry breaking term, such as (4.37) for example, will be added later. The fields S and P which appear in the quark loop term are the $N_f \times N_f$ matrices $S_a\tau_a$ and $P_a\tau_a$ with $a = 0, 1, 2, ..., N_f^2 - 1$. Using the identities:

$$S_a^2 = \frac{1}{2}tr_f S_a\tau_a S_b\tau_b = \frac{1}{2}tr_f S^2 \tag{5.2}$$

we rewrite the action (5.1) in the form:

$$I(S,P) = -Tr\ln\left(-i\partial_\mu\gamma_\mu + S + i\gamma_5 P\right) + \frac{1}{4g_s^2}tr_f\int d_4x\left((S-m)^2 + P^2\right) \tag{5.3}$$

where tr_f is a trace over the flavor indices. In writing (5.3), we have shifted the current quark mass term by making the substitution $m + S \to S$. The action (5.3) is a functional of the fields $S_{ij}(x)$ and $P_{ij}(x)$, which are matrices in flavor space.

Let us define the Dirac operator:

$$D = -i\partial_\mu\gamma_\mu + S + i\gamma_5 P \tag{5.4}$$

Calculations in the meson sector are simplified if we separate the quark loop term into real and imaginary parts:

$$-Tr\ln D = -\frac{1}{2}Tr\ln D^\dagger D + \frac{1}{2}Tr\ln D^\dagger/D \tag{5.5}$$

We can evaluate $D^\dagger D$ explicitly:

$$D^\dagger D = -\partial_\mu^2 + S^2 + P^2 + i\gamma_5\left[S,P\right] + i\gamma_\mu\left(\partial_\mu S\right) + \gamma_5\gamma_\mu\left(\partial_\mu P\right) \tag{5.6}$$

We will check a posteriori that a stationary point of the action exists which is translationally invariant and which is diagonal in flavor space:

$$S = M \equiv \sum_{i=u,d,s,...} |i\rangle M_i \langle i| \qquad P = 0 \qquad (\partial_\mu M) = 0 \tag{5.7}$$

At the point (5.7), the Dirac operator is $D = -i\partial_\mu\gamma_\mu + M$. It describes free $u, d, s, ...$ quarks of masses M_i which are usually called *constituent quark masses*, as opposed to the current quark masses m_i. The values of the fields at a stationary point of the action are often referred to as the *classical values of the fields*. We shall see that, except for Goldstone bosons, the masses of light hadrons turn out to be close to the masses of their constituent quarks.

We substitute the expression (5.6) into the action (5.3) and we redefine the action by subtracting its value at the point (5.7):

$$I(S,P) \to I(S,P) - I(S=M, P=0)$$

$$= -\frac{1}{2}Tr\ln(1+GV) + \frac{1}{4g_s^2}tr_f\int d_4x\left(S^2 + P^2 - M^2 - 2m(S-M)\right)$$

$$-\frac{1}{2}Tr\ln(1+gv) + \frac{1}{2}Tr\ln\left(1+gv^\dagger\right) \tag{5.8}$$

where we defined the propagators:

$$G \equiv \frac{1}{-\partial_\mu^2 + M^2} \qquad g = \frac{1}{-i\partial_\mu\gamma_\mu + M} \tag{5.9}$$

and:

$$V = S^2 + P^2 - M^2 + i\gamma_5\left[S,P\right] + i\gamma_\mu\left(\partial_\mu S\right) + \gamma_5\gamma_\mu\left(\partial_\mu P\right)$$

$$v = S - M + i\gamma_5 P \tag{5.10}$$

The last line of (5.8) is the contribution of the imaginary part $\frac{1}{2}Tr\ln D^\dagger/D$ of the action. To write it, we used the identity $\gamma_5 g^\dagger\gamma_5 = g$.

Consider the expansion of the action (5.8), about the point (5.7), in powers of V and v. The last line of (5.8) is odd in $i\gamma_5 P$. In order to survive the trace, four γ_μ Dirac matrices are required. They can only come from four quark propagators g. The imaginary part therefore begins only at *fifth* order in v. It is convergent and it gives rise to anomalous processes such as $KK \to 3\pi$. It does not contribute to any of the quantities discussed in this chapter. We therefore drop it.

5.2 The gap equations

Consider the variation $S \to S + \delta S$ and $P \to P + \delta P$ of the fields. The corresponding first order variation of the action (5.8), at the point (5.7), is

$$\delta I = -TrGM\delta S + \frac{1}{2g_s^2} tr_f \int d_4x \, (M - m) \, \delta S \tag{5.11}$$

No first order contribution in P occurs, which confirms that a stationary point with $P = 0$ does indeed exist. When a $U_A\,(1)$ symmetry breaking term of the form (4.37) is added to the action, it does not contribute to the first order variation of the action. We evaluate the traces appearing in the first term of (5.11) using the plane wave basis (3.31) in which G is diagonal:

$$G\,|k\rangle = \frac{1}{k^2 + M^2}\,|k\rangle \equiv G\left(k^2\right)|k\rangle \qquad \langle x \mid k\rangle = \frac{1}{\sqrt{\beta\Omega}} e^{ik_\mu x_\mu} \tag{5.12}$$

Performing the traces and remembering that M, G and m are diagonal in flavor space, the first order variation of the action becomes:

$$\delta I = -4N_c \frac{1}{\beta\Omega} \sum_{ki} \frac{M_i}{k^2 + M_i^2} \int d_4x \delta S_{(ii)} + \frac{1}{2g_s^2} \sum_i (M_i - m_i) \int d_4x \delta S_{(ii)} \tag{5.13}$$

where the flavor indices $i = u, d, s, ...$ run over the N_f flavor variables and where $\delta S_{(ii)}$ is the diagonal matrix element $\langle i\,|\delta S|\,i\rangle$. No first order off-diagonal matrix elements δS_{ij} occur, nor do any gradients of S. This confirms the existence of a stationary point of the form (5.7). To determine the stationary point, we set to zero the terms which are first order in $S_{(ii)}$:

$$\frac{1}{8N_c g_s^2}\left(1 - \frac{m_i}{M_i}\right) = \frac{1}{\beta\Omega} \sum_{k<\Lambda} \frac{1}{k^2 + M_i^2} \tag{5.14}$$

A quadratically divergent sum over k occurs. Various regularizations of the fermion determinant $-Tr \ln D$ are discussed in chapter 6. At this point we shall simply assume that the 4-momenta are cut off at some value $\left|k_\mu^2\right| < \Lambda$. The value of the cut-off Λ is estimated in section 5.5. The equations (5.14) are called the *gap equations* because the constituent quark masses M_i determine the energy gap separating the positive and negative energy quark orbits. When non-local and proper-time regularizations are used, the gap equations acquire the forms (9.5) and (9.15) respectively.

The constituent quark mass M, discussed in this chapter, includes the current quark mass m. This is because of the substitution $m + S \rightarrow S$ made above in order to write the action in the form (5.3).

The gap equations determine the constituent quark masses M_i which make the action (5.8) stationary for any given value of the coupling constant g_s^2. We can however use them the other way around, namely to eliminate the coupling constant g_s^2 in favor of the constituent quark masses M_i. To do this, we note that the propagator (5.12) is diagonal in flavor space, so that:

$$\begin{aligned} TrGV &= TrG\left(S^2 + P^2 - M^2\right) \\ &= 4N_c \sum_i \left(\frac{1}{\beta\Omega} \sum_{k<\Lambda} \frac{1}{k^2 + M_i^2} \right) \int d_4x \left\langle i \left| S^2 + P^2 - M^2 \right| i \right\rangle \end{aligned} \quad (5.15)$$

We can then use the gap equations (5.14) to deduce the equality:

$$\frac{1}{2} TrGV = \frac{1}{4g_s^2} tr_f \int d_4x \left(1 - \frac{m}{M}\right) \left(S^2 + P^2 - M^2\right) \quad (5.16)$$

When this result is inserted into the expression (5.8), we obtain the action in the form:

$$I(S,P) = -\frac{1}{2} Tr \ln(1 + GV) + \frac{1}{2} TrGV + \frac{1}{4g_s^2} tr_f \frac{m}{M} \int d_4x \left((S - M)^2 + P^2 \right) \quad (5.17)$$

The form (5.17) is very convenient for expansions of the action about the stationary point. Indeed, since V is first order in the fluctuating parts of the fields, the action (5.17) is explicitly stationary with respect to variations of the fields S and P. Furthermore the second term $\frac{1}{2}TrGV$ cancels the quadratic divergence of the first term, leaving only a logarithmic divergence.

The equivalent sigma model is obtained by retaining the most divergent piece of the action (5.17). Since V is a local interaction, the only divergent part is second order in V. Let us expand the action (5.17) to second order in V :

$$I^{(2)}(S,P) = \frac{1}{4} TrGVGV + \frac{1}{4g_s^2} tr_f \frac{m}{M} \int d_4x \left((S - M)^2 + P^2 \right) \quad (5.18)$$

Higher order terms are convergent. The divergent part of $TrGVGV$ is equal to TrG^2V^2 as shown by the identity:

$$TrGVGV = TrG^2V^2 + \frac{1}{2} TrG^2 \left[G^{-1}, V\right] G^2 \left[G^{-1}, V\right] \quad (5.19)$$

The commutator $\left[G^{-1}, V\right]$ is of order k whilst G^2 is of order k^{-4}. Therefore the second term of the identity (5.19) is of order k^{-6}, whereas the first term diverges logarithmically. Thus, retaining only the most divergent part, we obtain the approximate action:

$$I_{\sigma-model}(S,P) = \frac{1}{4}TrG^2V^2 + \frac{1}{4g_s^2}tr_f\frac{m}{M}\int d_4x\left((S-M)^2+P^2\right) \tag{5.20}$$

which we refer to as the action of the *equivalent linear sigma model* (see section 5.4). The logarithmically divergent action (5.20) contains the leading order parts of the action (5.17) when the cut-off Λ is much larger than the constituent quark masses.

In the following we shall assume that the current masses of the u and d quarks are equal:

$$m_u = m_d \equiv m \neq m_s, ... \tag{5.21}$$

This makes the Nambu Jona-Lasinio action invariant under isospin rotations (2.2) and it leads to hadrons which form degenerate isospin multiplets. Effects due to the small difference between m_u and m_d are no larger than effects due to electromagnetic interactions, which we have also neglected [48]. The gap equations show that when the condition (5.21) is satisfied, a similar relation holds for the constituent quark masses:

$$M_u = M_d \equiv M \neq M_s, ... \tag{5.22}$$

No confusion should arise from the use the symbol M to denote either the matrix M or the constituent quark masses $M_u = M_d$.

5.3 The large N_c limit and caveats thereon

The quark loop term $-Tr \ln D$ of the action involves a trace over color-independent operators. It is therefore proportional to the number of colors N_c. The action (5.1) becomes proportional to N_c provided that we assume that the coupling constant g_s^2 is inversely proportional to N_c. The N_c counting rules of QCD [60, 61] assume that the QCD coupling constant α^2 is inversely proportional to N_c. Both the gluon exchange and the instanton induced interactions discussed in chapter 3 are inversely proportional to N_c. It is perhaps reasonable to assume that the coupling constants g_s^2 and g_v^2 of the quark interactions should also be inversely proportional to N_c. The gap equation (5.14) then ensures that the constituent quark mass M is independent of N_c. The N_c counting rules of QCD are then respected by the model: the masses of baryons (which are composed of N_c quarks) are proportional to N_c; the masses of mesons (which are $q\bar{q}$ excitations) are proportional to $(N_c)^0$; interactions between mesons are proportional to $1/N_c$ etc.

It has been argued that, in the large N_c limit, QCD reduces to an effective theory of non-interacting mesons, and that baryons emerge as solitons [61]. This argument is often abusively invoked to justify Skyrme and Nambu Jona-Lasinio models. Indeed, the argument does not specify the form of the low energy effective action, so that it cannot be used to claim that specific models, such as Nambu Jona-Lasinio or Skyrme models, are derived from QCD. One should also remember that it has never yet been proved that confinement is maintained in the large N_c limit.

Consideration of the large N_c limit is however useful because, when an action is proportional to N_c, it provides a systematic expansion parameter even if the interactions are strong. To illustrate this, consider a euclidean action $N_c I(S)$ of a single field S. The stationary point $S = M$ of the action satisfies the equation $\delta I/\delta S|_{S=M} = 0$ and it is independent of N_c. We can then evaluate the partition function using the saddle point method:

$$Tre^{-\beta H} \equiv e^{-W} = \int D(S)\, e^{-N_c I(S)} = e^{-N_c I(M)} \int D(S)\, e^{-\frac{1}{2}N_c S \left.\frac{\delta^2 I}{\delta S \delta S}\right|_{S=M} S} \tag{5.23}$$

where W is the energy of the system. Performing the gaussian integral, we obtain:

$$W = N_c I(M) + \frac{1}{2} Tr \ln \left.\frac{\delta^2 I}{\delta S \delta S}\right|_{S=M} + \mathcal{O}\left(\frac{1}{N_c}\right)$$

The neglected term $\frac{1}{2} Tr \ln N_c$ is dropped because it simply redefines the zero energy. The leading order term is the *classical energy*, which is equal to the euclidean action evaluated at the stationary point. It is of order N_c. The second term, which is called the *one loop* term, is the contribution of the quantum fluctuations of the field S about the stationary point $S = M$. It is proportional to $(N_c)^0$, down by a factor $1/N_c$ relative to the first.

Quarks have only $N_c = 3$ colors. This is not enough to ensure that successive terms of the loop expansion decrease for all physical quantities. For example, the low temperature dependence of the vacuum quark condensate is dominated by thermal excitations of pions. The latter should be down by a factor $1/N_c$ relative to quark excitations. However, the low mass of the pions, relative to the constituent quark mass, more than compensates the $1/N_c$ reduction factor (see section 10.5.2). Some $1/N_c$ corrections are not negligible. For example, the center-of-mass correction of solitons, which are bound states of N_c quarks, amounts to a roughly $1/N_c \approx 30\%$ correction of the nucleon mass. The axial coupling constant g_A is enhanced by almost 40% by its coupling to the soliton rotation [62].

5.4 The equivalent linear sigma model in a simplified $SU(2)_f$ case

We consider a simplified version of the Nambu Jona-Lasinio model in which only u and d quarks are included, so that $N_f = 2$. We include only a singlet isoscalar scalar field S (which will be proportional to the σ meson field) and a triplet of isovector pseudoscalar fields $P_{a=1,2,3}$ (which will be proportional to the three π meson fields). The Nambu Jona-Lasinio action (4.23) reduces to:

$$I(S,P) = -Tr \ln \left(-i\partial_\mu \gamma_\mu + m + S + i\gamma_5 P_a \tau_a\right) + \frac{1}{2g_s^2} \int d_4x \left(S^2 + P_a^2\right) \tag{5.24}$$

We assume equal masses for the u and d quarks. The simplified action (5.24) is invariant under isospin rotations (2.2) and chiral transformations (2.3). It is not invariant under axial $U_A(1)$ transformations (2.4).

In this simple model, the action (5.20) of the equivalent sigma model reduces to:

$$I_{\sigma-model}(S,P) = \frac{1}{4} TrG^2V^2 + \frac{m}{2g_s^2 M} \int d_4x \left((S-M)^2 + P_a^2 \right) \tag{5.25}$$

and the expression (5.10) for V reduces to:

$$V = S^2 + P^2 - M^2 + i\gamma_\mu (\partial_\mu S) + \gamma_5 \gamma_\mu (\partial_\mu P) \tag{5.26}$$

with $P^2 = P_a^2$. We take the trace using the plane wave basis (3.31). Thus, for example:

$$\frac{1}{4} TrG^2V^2 = \frac{1}{4} TrG^2 \left(\left(S^2 + P_a^2 - M^2\right)^2 + (\partial_\mu S)^2 + (\partial_\mu P_a)^2 \right)$$

$$= \frac{1}{4} 8N_c \left(\frac{1}{\beta\Omega} \sum_{k<\Lambda} \frac{1}{(k^2+M^2)^2} \right) \int d_4x \left(\left(S^2 + P_a^2 - M^2\right)^2 + (\partial_\mu S)^2 + (\partial_\mu P_a)^2 \right) \tag{5.27}$$

The factor $8N_c$ is due to the trace over the Dirac, flavor and color indices. Proceeding in this way, we reduce the action (5.20) to:

$$I_{\sigma-model}(S,P) = \int d_4x \mathcal{L}_{\sigma-model}(S(x), P(x))$$

$$\mathcal{L}_{\sigma-model}(S,P) = \frac{1}{2} Z_\pi \left((\partial_\mu S)^2 + (\partial_\mu P_a)^2 + \left(S^2 + P_a^2 - M^2\right)^2 \right)$$

$$+ \frac{m}{2g_s^2 M} \left((S-M)^2 + P_a^2 \right) \tag{5.28}$$

where:

$$Z_\pi = 4N_c \frac{1}{\beta\Omega} \sum_{k<\Lambda} \frac{1}{(k^2+M^2)^2} \tag{5.29}$$

We have simply cut off the logarithmically divergent sum over the 4-momenta k at a value $k^2 < \Lambda^2$. A more detailed discussion of regularization schemes is given in chapter 6. We define σ and π meson fields so as to normalize their kinetic energies:

$$\sigma(x) = \sqrt{Z_\pi} S \qquad \pi_a(x) = \sqrt{Z_\pi} P_a(x) \qquad f_\pi = \sqrt{Z_\pi} M \tag{5.30}$$

and we define the pion mass:

$$m_\pi^2 = \frac{m}{Z_\pi g_s^2 M} = \frac{mM}{f_\pi^2 g_s^2} \tag{5.31}$$

The lagrangian (5.28) acquires the familiar form of a linear sigma model:

$$\mathcal{L}_{\sigma-model}(\sigma,\pi) = \frac{1}{2}(\partial_\mu \sigma)^2 + \frac{1}{2}(\partial_\mu \pi_a)^2 + \frac{M^2}{2f_\pi^2}\left(\sigma^2 + \pi_a^2 - f_\pi^2\right)^2$$
$$+\frac{m_\pi^2}{2}\left((\sigma - f_\pi)^2 + \pi_a^2\right) \tag{5.32}$$

The mass of the σ-meson is the coefficient of the term proportional to $(\sigma - f_\pi)^2/2$:

$$m_\sigma^2 = 4M^2 + m_\pi^2 \tag{5.33}$$

The quark condensate is:

$$\langle \bar{q}(x) q(x) \rangle = \left\langle \bar{u}(x) u(x) + \bar{d}(x) d(x) \right\rangle = \frac{1}{\beta\Omega}\frac{\delta I}{\delta m} \tag{5.34}$$

where the current quark mass was used as a source term. Using the form (5.1) of the action and the definition (5.31) of the pion mass, we obtain, in the chiral limit:

$$\langle \bar{q} q \rangle = -\frac{M-m}{g_s^2} \approx -\frac{f_\pi^2 m_\pi^2}{m} \tag{5.35}$$

a result known as the Gell-Mann Oakes Renner formula [63].

5.5 Setting the scale by fitting f_π

It is customary to set the scale for all the quantities by fitting the observed pion decay constant $f_\pi = 93$ MeV. Using (5.29) and (5.30), we obtain:

$$f_\pi^2 = 4N_c M^2 \frac{1}{\beta\Omega}\sum_{k<\Lambda}\left(\frac{1}{k^2+M^2}\right)^2 = 4N_c M^2 \frac{\pi^2}{(2\pi)^4}\int_0^{\Lambda^2} y dy \left(\frac{1}{y+M^2}\right)^2 \tag{5.36}$$

This equation relates the constituent quark mass M to the cut-off Λ. It can be cast into the more transparent form:

$$\frac{16\pi^2}{4N_c}\frac{f_\pi^2}{\Lambda^2} = \frac{M^2}{\Lambda^2}\int_0^1 \frac{x dx}{\left(x + \frac{M^2}{\Lambda^2}\right)^2} \tag{5.37}$$

The graphical solution is shown in Fig.5.1:

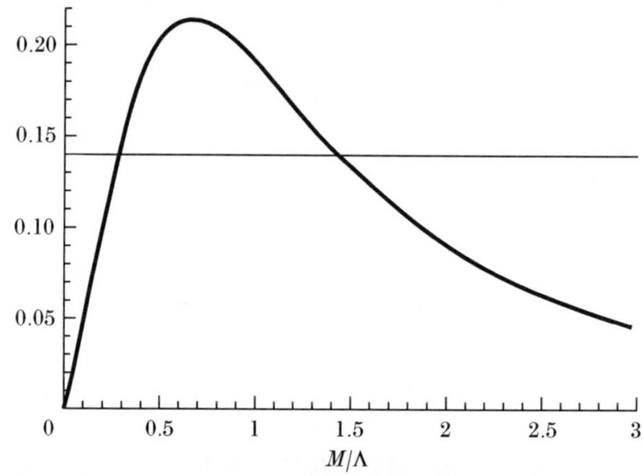

FIG. 5.1. Graphical solution of eqn (5.37). The right hand side is plotted as a function of M/Λ

The curve reaches a maximum equal to 0.216 for $\frac{M}{\Lambda} = 0.68$. We deduce that the cut-off has a lower limit:

$$\Lambda^2 > 4.6 \, \frac{16\pi^2}{4N_c} \, f_\pi^2 \tag{5.38}$$

For $N_c = 3$ and $f_\pi = 93$ MeV this condition becomes $\Lambda > 0.72$ GeV. Table 5.1 shows the values of the cut-off Λ and M which are related by the condition (5.36).

We can use the gap equation (5.14) to evaluate the coupling constant g_s^2 in the chiral limit $m \to 0$. The quark condensate can then be deduced from the expression (5.35) and the cut-off from (5.37). The quark condensate thus also acquires a lower limit:

$$\left| \langle \bar{q}q \rangle^{1/3} \right| > 3.4 \, \frac{f_\pi}{(N_c)^{1/6}} \tag{5.39}$$

For $N_c = 3$ and $f_\pi = 93$ MeV, a minimum $\left| \langle \bar{q}q \rangle^{1/3} \right| = 0.26$ GeV occurs for $\frac{M}{\Lambda} = 0.64$. Table 5.1 shows the values of various quantities when the scale is set by fitting f_π:

The values quoted in Table 5.1 have been obtained by simply cutting off the 4-momenta at $k^2 = \Lambda^2$. Other regularizations can yield somewhat different numbers. For a large cut-off, the relation (5.37) between M and Λ depends only logarithmically on the cut-off. Quantities which depend logarithmically on the cut-off are not very dependent on which regularization is used, although they do depend quite strongly on the value of the cut-off. Quantities such as the quark condensate $\langle \bar{q}q \rangle$, the current quark mass m and the coupling constant g_s^2 depend

Table 5.1 *Values of quantities with the scale set by fitting f_π*

$= \frac{M}{\Lambda}$	0.1	0.2	0.4	0.6	0.8	1.0	1.2	1.4
(GeV)	1.77	1.11	0.80	0.73	0.73	0.76	0.82	0.88
M (GeV)	0.18	0.22	0.32	0.44	0.58	0.76	0.98	1.23
$\frac{g_\sigma^2 \Lambda^2}{4\pi}$	0.55	0.60	0.77	1.00	1.31	1.71	2.18	2.72
$\langle \bar{q}q\rangle^{1/3}$ (GeV)	−0.432	−0.331	−0.276	−0.264	−0.267	−0.276	−0.289	−0.304

quadratically on the cut-off, as illustrated in the Table 10.2, of chapter 10. They depend far more on which regularization is used. It has been noted that they are also the ones which, in lattice calculations, have an anomalous dimension [64].

The quark condensate has been estimated from QCD sum rules [43] to be $\langle \bar{u}u\rangle^{1/3} \simeq \langle \bar{d}d\rangle^{1/3} \simeq -225$ MeV at the scale of 1 GeV, which corresponds to $\langle \bar{q}q\rangle^{1/3} \simeq -283$ MeV. We see from Table 5.1 that this value is close to the minimum value (5.39) but that it does not constrain the constituent quark mass M very seriously. The sharp four-momentum cut-off used here has a tendency to reduce quark condensates, which depend quadratically on the cut-off. The QCD estimate however is model-dependent. It has been argued that considerably lower values of the quark condensate are compatible with the presently available data [65].

A striking feature of Table 5.1 is the low value of the cut-off. The ratio $\frac{M}{\Lambda}$ is small and the cut-off Λ large only for low values of M. However, fitting hadron masses requires constituent quark masses $M \approx 0.35 - 0.5$ GeV or higher. Indeed, soliton energies, which represent baryon masses, are of the order of $N_c M$ and meson masses (except for Goldstone bosons) are of order $2M$. A possible candidate for the σ meson could be the $J^{PC} = 0^{++}$ meson $f_0\,(980)$. From the equation (5.33) we see that $4M^2 = m_{f_0}^2 - m_\pi^2$. Using $m_{f_0} = 980$ MeV and $m_\pi = 138$ MeV, we find $M = 485$ MeV . This is more thoroughly discussed in section 5.6.2. Solitons, which are bound states of quarks, occur only for $M \gtrsim 400$ MeV (see section 12.8). From Table 5.1 we see that large constituent quark masses correspond to strong coupling constants g_s^2. The coupling constant needs to be strong enough to allow the chiral field to bind the quarks. This is further discussed in section 5.6.2. One can define a dimensionless quark–meson coupling constant g, by assuming that the constituent quark mass is M is due to the coupling of the quark to the vacuum value $\sigma = f_\pi$ of the sigma meson field. Thus

$$M = g f_\pi \tag{5.40}$$

For constituent quark masses $M \approx 400$ MeV, the coupling constant is $g \approx 4$, which is close to the values required to bind solitons in the Gell-Mann Lévy model (section 11.3).

5.6 $SU(3)_f$ scalar fields

We turn to the case of $N_f = 3$ flavors. The first term of the equivalent sigma model action (5.20) can be calculated on the plane wave basis (3.31):

$$\frac{1}{4} TrG^2V^2 = \frac{1}{4}\frac{4N_c}{\beta\Omega}\sum_{k<\Lambda}\frac{1}{(k^2+M^2)^2} tr_f \int d_4x V^2 \tag{5.41}$$

where tr is a trace over the flavor indices.. The factor $4N_c$ is due to the trace over the Dirac and color indices.. We have:

$$tr_f \frac{1}{(k^2+M^2)^2} V^2 = tr_f \frac{1}{(k^2+M^2)^2} \left(\left(S^2+P^2-M^2\right)^2 - [S,P]^2 + (\partial_\mu S)^2 + (\partial_\mu P)^2\right) \tag{5.42}$$

The explicit form of the equivalent sigma model action is somewhat complicated in the case of $SU(3)_f$. We therefore simply write the action up to second order in the fluctuating parts $\widetilde{S} = S - M$ and P of the fields. To this order, (5.42) reduces to:

$$tr \frac{1}{(k^2+M^2)^2} V^2 = 4N_c \sum_{ij} \frac{1}{(k^2+M_i^2)^2} \left(\left(\partial_\mu \widetilde{S}_{ij}\right)\left(\partial_\mu \widetilde{S}_{ji}\right) + (\partial_\mu P_{ij})(\partial_\mu P_{ji}) \right.$$

$$\left. (M_i+M_j)^2 \widetilde{S}_{ij}\widetilde{S}_{ji} + (M_i-M_j)^2 P_{ij}P_{ji}\right) \tag{5.43}$$

where $i, j = u, d, s$ are the flavor indices.. The action of the equivalent linear sigma model (5.20) can therefore be reduced to the form:

$$I_{\sigma-model} = \int d_4x \left(\mathcal{L}_s + \mathcal{L}_{ps}\right) \tag{5.44}$$

where the scalar meson term is:

$$\mathcal{L}_s = \sum_{ij} \frac{1}{4} Z_{ij}\left(\left(\partial_\mu \widetilde{S}_{ij}\right)\left(\partial_\mu \widetilde{S}_{ji}\right) + (M_i+M_j)^2 \widetilde{S}_{ij}\widetilde{S}_{ji}\right) + \frac{1}{8g_s^2}\left(\frac{m_i}{M_i} + \frac{m_j}{M_j}\right)\widetilde{S}_{ij}\widetilde{S}_{ji} \tag{5.45}$$

and the pseudoscalar term is:

$$\mathcal{L}_{ps} = \sum_{ij} \frac{1}{4} Z_{ij}\left((\partial_\mu P_{ij})(\partial_\mu P_{ji}) + (M_i-M_j)^2 P_{ij}P_{ji}\right) + \frac{1}{8g_s^2}\left(\frac{m_i}{M_i} + \frac{m_j}{M_j}\right)P_{ij}P_{ji} \tag{5.46}$$

The function Z_{ij} is:

$$Z_{ij} = \frac{2N_c}{\beta\Omega}\sum_{k<\Lambda}\left(\frac{1}{(k^2+M_i^2)^2} + \frac{1}{(k^2+M_j^2)^2}\right) \tag{5.47}$$

which we have regularized by simply cutting off the 4-momentum sum. A $U_A(1)$ symmetry breaking term (5.52) will be added below to the pseudoscalar term $\mathcal{L}_{ps}$.

The matrix S in flavor space can be expanded on the basis of the $U(3)$ matrices (A.22), which include the unit matrix $\tau_0 = \sqrt{\frac{2}{3}}$. Its explicit representation is then:

$$S \equiv S_a \tau_a = \begin{pmatrix} S_3 + S_0' & S_1 - iS_2 & S_4 - iS_5 \\ S_1 + iS_2 & -S_3 + S_0' & S_6 - iS_7 \\ S_4 + iS_5 & S_6 + iS_7 & -\sqrt{2} S_8' \end{pmatrix} \tag{5.48}$$

where:

$$S_0' = \sqrt{\frac{2}{3}} S_0 + \frac{1}{\sqrt{3}} S_8 \qquad S_8' = -\frac{1}{\sqrt{3}} S_0 + \sqrt{\frac{2}{3}} S_8 \tag{5.49}$$

The linear combinations (5.49) are commonly expressed in terms of the so-called *ideal mixing angle* $\theta = 0.6155 = 35.26°$, which is such that $\cos\theta = \sqrt{\frac{2}{3}}$. A form similar to (5.48) applies to the matrix P_{ij}. Operators such as (5.7) are diagonal in flavor space. They can be written in the form:

$$M = M_0 \tau_0 + M_3 \tau_3 + M_8 \tau_8$$

$$M_0 = \frac{M_u + M_d + M_s}{\sqrt{6}} \qquad M_3 = \frac{M_u - M_d}{2} \qquad M_8 = \frac{M_u + M_d - 2M_s}{2\sqrt{3}} \tag{5.50}$$

We assume, as in (5.21) and (5.22), that the matrices m, M and G commute with τ_3 and we use the notation:

$$m_u = m_d \equiv m \neq m_s \qquad G_u = G_d \equiv G \neq G_s \qquad M_u = M_d \equiv M \neq M_s \tag{5.51}$$

5.6.1 *The pseudoscalar* π, K, η *nonet* $(L = 0, \ J^P = 0^-)$

We need to add a $U_A(1)$ symmetry breaking term to the lagrangian (5.46), otherwise we cannot account for the observed masses of the $\eta(547)$ and $\eta'(958)$ mesons. For simplicity, we adopt the Shechter form (4.37) which we expand to second order in the fluctuating parts of the fields:

$$V_{ven} = -\frac{\kappa^2}{2} \left(tr_f \ln(S + iP) - tr_f \ln(S - iP) \right)^2 = 2\kappa^2 \left(tr_f \frac{1}{M} P \right)^2$$

$$= 2\kappa^2 \left(\sum_{i=u,d,s} \frac{1}{M_i} P_{ii} \right)^2 = 4\kappa^2 \left(\frac{\sqrt{2}}{M} P_0' - \frac{1}{M_s} P_8' \right)^2 \tag{5.52}$$

The symmetry breaking term (5.52) does not modify the gap equations. It involves only the pseudoscalar fields P_0' and P_8' fields and not the scalar fields S_0'

and $S_0^{'}$. This is why the ideal mixing angle remains unperturbed for the scalar fields. If we explicit the matrix P_{ij} using (5.48), the lagrangian (5.46) reads:

$$\mathcal{L}_{ps} = Z_\pi \left(\frac{1}{2} \left(\partial_\mu P_1\right)^2 + ... + \frac{1}{2} \left(\partial_\mu P_3\right)^2 + \frac{m_\pi^2}{2} \left(P_1^2 + ... + P_3^2\right) \right)$$

$$+ Z_K \left(\frac{1}{2} \left(\partial_\mu P_4\right)^2 + ... + \frac{1}{2} \left(\partial_\mu P_7\right)^2 + \frac{m_K^2}{2} \left(P_4^2 + ... + P_7^2\right) \right)$$

$$+ Z_\pi \left(\frac{1}{2} \left(\partial_\mu P_0^{'}\right)^2 + \frac{m_\pi^2}{2} \left(P_0^{'2}\right) \right) + Z_\eta \left(\frac{1}{2} \left(\partial_\mu P_8^{'}\right)^2 + \frac{m_\eta^2}{2} P_8^{'2} \right)$$

$$+ 4\kappa^2 \left(\frac{\sqrt{2}}{M} P_0^{'} - \frac{1}{M_s} P_8^{'} \right)^2 \tag{5.53}$$

where:

$$Z_\pi = Z_{uu} \qquad Z_K = Z_{us} \qquad Z_\eta = Z_{ss} \tag{5.54}$$

with Z_{ij} given by (5.47). The masses are equal to:

$$m_\pi^2 = \frac{m}{Z_\pi g_s^2 M} \quad m_K^2 = \frac{1}{2 Z_K g_s^2} \left(\frac{m}{M} + \frac{m_s}{M_s} \right) + \left(M - M_s\right)^2 \quad m_\eta^2 = \frac{m_s}{Z_\eta g_s^2 M_s} \tag{5.55}$$

The three fields $\sqrt{Z_\pi} P_{a=1,2,3}$ are associated to the three pions $\pi\,(138)$ and the four fields $\sqrt{Z_K} P_{a=4,5,6,7}$ are associated to the four kaons $K\,(495)$, shown in Fig. 5.2. Were it not for the last term of (5.53), which is due to the $U_A\,(1)$ symmetry breaking term (4.37), the fields $\sqrt{Z_\pi} P_0^{'}$ and $\sqrt{Z_\eta} P_8^{'}$ would be associated to the $\eta'\,(958)$ and $\eta\,(547)$ mesons respectively and the η' would be degenerate with the pion. In order to determine the physical mesons, the mass terms involving $P_0^{'}$ and $P_8^{'}$ need to be diagonalized. We shall not do so here. Results obtained by various groups are summarized in section 10.1. From the equation (5.55) we see that the experimental pion and kaon masses m_π and m_K can be fitted by an appropriate choice of the *current* quark masses m and m_s. This can also be expressed in terms of Gell-Mann Oakes Renner relations [63]. The strength κ^2 of the $U_A\,(1)$ symmetry breaking term can be fitted to the observed mass of the $\eta'\,(958)$. The mass of the η then comes out within about 15% of the observed mass of the $\eta\,(547)$. The masses of the pseudoscalar nonet π, K, η are not determined by the coupling constant g_s^2 nor by the constituent quark masses M_i. Indeed, in the chiral limit $m \to 0$, the π, K and η would have zero mass, independently of the coupling constant. This is a result of the spontaneously broken symmetry. Thus, in the Nambu Jona-Lasinio model, it is the spontaneously broken chiral symmetry which binds $q\bar{q}$ excitations of the pseudoscalar nonet, rather than the particular form or the strength of the quark–quark interaction.The weak decay constants f_π, f_K and the quark condensates have almost the same values as those calculated more rigorously from correlation functions (see sections 9.2.2 and 9.2.3).

5.6.2 *The scalar a_0, K^*, f_0 nonet* $(L = 1,\ J^P = 0^+)$

If we explicit the field S_{ij} according to (5.48), the lagrangian of the scalar part (5.45) of the equivalent sigma model becomes:

$$\mathcal{L}_\sigma = \frac{Z_\pi}{2}\left(\left(\partial_\mu \widetilde{S}_1\right)^2 + \ldots + \left(\partial_\mu \widetilde{S}_3'\right)^2 + \left(4M^2 + m_\pi^2\right)\left(\widetilde{S}_1^2 + \ldots + \widetilde{S}_3'^2\right)\right)$$
$$+\frac{Z_K}{2}\left(\left(\partial_\mu \widetilde{S}_4\right)^2 + \ldots + \left(\partial_\mu \widetilde{S}_7\right)^2 + \left(4MM_s + m_K^2\right)\left(\widetilde{S}_4^2 + \ldots + \widetilde{S}_7^2\right)\right)$$
$$+\frac{Z_\pi}{2}\left(\left(\partial_\mu \widetilde{S}_0'\right)^2 + \left(4M^2 + m_\pi^2\right)\widetilde{S}_0'^2\right) + \frac{Z_\eta}{2}\left(\left(\partial_\mu \widetilde{S}_8'\right)^2 + \left(4M_s^2 + m_\eta^2\right)\widetilde{S}_8'^2\right) \tag{5.56}$$

The isoscalar nonet is interesting because it provides, in principle, a direct determination of the constituent quark masses. Assume that we assign the nine properly normalized fields $\widetilde{S}_a$ to the nonet of $J^P = 0^+$ mesons, shown in Figure 5.2.

The three fields $\sqrt{Z_\pi}\widetilde{S}_{a-1,2,3}$ arc then associated to the isotriplet meson $a_0\,(980)$. The field $\sqrt{Z_\pi}\widetilde{S}_0'$ is associated to the meson $f_0\,(980)$. The four fields $\sqrt{Z_K}\widetilde{S}_{a=4,5,6,7}$ are associated to the $K_0^*\,(1430)$, and the field $\sqrt{Z_\eta}\widetilde{S}_8'$ to the $f_0'\,(1300)$. These assignments are taken from Table 30.2 of the Particle Data [7] and they are further discussed below. The masses can be read off (5.56):

$$M_{a_0}^2 = M_{f_0}^2 = 4M^2 + m_\pi^2 \qquad M_{K_0^*}^2 = 4MM_s + m_K^2 \qquad M_{f_0'}^2 = 4M_s^2 + m_\eta^2 \tag{5.57}$$

(As noted above, the m_η is not the observed η meson mass.). The observed mass of the $a_0\,(980)$ implies a constituent u and d quark mass $M = 485$ MeV, considerably higher than the values usually adopted, quoted in Table 10.1. The meson $f_0\,(980)$ has the same mass as the $a_0\,(980)$, in agreement with the lagrangian (5.56). If we determined the strange constituent quark mass M_s from the observed $K_0^*\,(1430)$ mass, we would obtain the estimate $M_s = 930$ MeV. However, this value would set the mass of the $f_0'\,(1300)$ above 1800 MeV. A lower mass $M_s \approx 700$ MeV would bring the masses of the K_0^* and the f_0' reasonably close to their experimental values. However, even this value is considerably higher than the usually adopted values. The assignment of $q\bar{q}$ configurations to scalar mesons is still a subject of considerable debate. For example, it has been suggested to assign the $a_0\,(1450)$, $K_0^*\,(1430)$, $f_0\,(1350)$ and $f_0\,(1525)$ or $f_0\,(1590)$ resonances

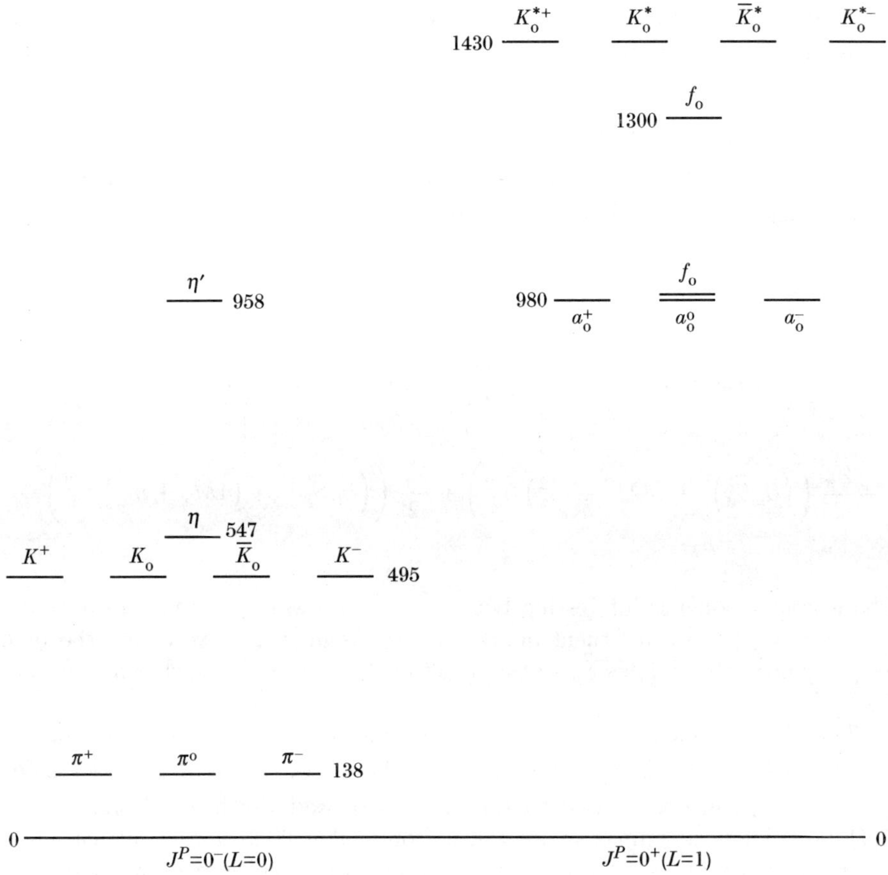

FIG. 5.2. Spectrum of the pseudoscalar and scalar meson nonets. The masses are taken from the 1994 Particle Data [7]

to the scalar $q\bar{q}$ nonet [66, 67]. These assignments would require even higher constituent quark masses $M \approx 650$ MeV. Although some fits to meson spectra, involving extra dilaton fields (section 11.2), do require high masses of the order of 500 MeV or higher [55], most applications so far have used lower constituent quark masses. This can be seen in Table 10.1 which lists parameters used by various groups which calculate on-shell meson masses. One should also remember that the sizes of solitons tend to shrink when they are calculated with high values of M, as illustrated in Table 12.1. But when constituent quark masses $M \approx 400$ MeV are used, the calculated K_0^* (1430) mass falls well below the observed value, as might be expected from the simple model estimate made above. Either the Nambu Jona-Lasinio model can accommodate higher constituent quark masses, or we must admit that it fails to bind the quarks which form the scalar meson nonet. This is further discussed in section 9.2.4. One may conjecture that the rather high constituent quark masses required to fit the hadron spectra are a consequence of the lack of confinement of the model. If additional confining forces were acting, smaller values of M might suffice to fit the scalar nonet masses. This is suggested by the results of constituent quark models which use confining interactions [1, 3]. They systematically get the scalar nonet at energies which exceed twice the constituent quark masses. This is understandable since, in the scalar nonet, the quark and antiquark have a relative orbital angular momentum $L = 1$. The wavefunction therefore extends further out into regions where the confining forces begin to act. The scalar nonet is probably an example of system in which confining forces begin to take over the chiral fields, to bind the quarks.

6

REGULARIZATION

The regularization of an effective theory is not just a mathematical trick designed to eliminate whatever infinities happen to occur. It is a physical process and, as such, an integral part of a model. If we knew how to derive a low energy effective theory from QCD, we would know how the theory becomes regularized. For example, the instanton induced interaction (section 3.2) involves delocalized fields, which introduce a non-locality which in turn regularizes the effective theory. More often than not, we adopt a given regularization scheme for convenience. This is tantamount to admitting our ignorance as to its origin. Regularization of an effective theory should also be distinguished from renormalization. A renormalizable model has the attractive feature that it can predict observables which do not depend on the cut-off, which can be taken to infinity. Many of the unpleasant properties encountered when an effective theory is regularized disappear when the cut-off is sent to infinity. However, the cut-off required to regularize the quark loop in the Nambu Jona-Lasinio model, for example, is quite low, of the order of 1 GeV, hardly exceeding the excitation energy of a $q\bar{q}$ pair. This precludes a useful hierarchy of energy scales. A model applicable only at energies which are an order of magnitude smaller than the cut-off is too restrictive.

The ultraviolet divergences which we have encountered so far relate to the quark loop term. Speculations vary as to the origin of the cut-off which could regulate this term. They appeal to the finite instanton size, to the range of non-locality of the auxiliary meson fields or to the asymptotic freedom of QCD.

Practically all regularizations in which the cut-off remains finite lead to unphysical singularities of the hadron propagators. This is a well-known feature of field theories which are regularized by non-local fields. But it also occurs in other regularizations, such as proper-time and Pauli–Villars regularizations. The occurrence of unphysical singularities can be invoked as a sufficient reason to exclude a fundamental field theory. It may not be a sufficient reason to exclude a model, however. We need to learn how to deal with effective theories and models which involve low cut-offs.

For the purpose of calculating leading order meson correlation functions and solitons, it is usually sufficient to regularize the fermion determinant $-Tr\ln D$. However, this does not always eliminate ultraviolet divergences which arise when higher order loops are included in the calculation of the partition function (4.19). In proper-time regularization, for example, the next to leading order meson loop contribution diverges and requires a further regularization. We shall see however that some non-local regularizations do regularize the model completely. The regularization of a fermion determinant is a subject which has been extensively studied (see [68] for a review). In this chapter, we limit the discussion to regularizations of models which describe quarks bound by chiral fields.

Two approaches towards regularization seem to have emerged. In the first approach, a regularized model is first defined, usually in terms of a regularized action. Physical observables and processes are derived from there. This approach includes, for example, non-local regularizations as well as proper-time and Pauli–Villars regularizations of the real part of the action. In the second approach, expressions for observables (typically correlation functions, quark condensates, etc.) are first formally derived from an unregularized model and divergences are then regularized wherever they turn up. This approach is popular because it leads to simpler calculations. However, it lacks the backing of a model. Correlation functions obtained as functional derivatives may not be consistent with the existence of an action. Regularization of different quantities can appear unrelated and the regularization can induce hidden features, such as non-locality, which can pass unnoticed.

The regularizations we proceed to discuss fall into two classes: non-local regularizations which regularize the full action, and regularizations which regularize only the real part $-\frac{1}{2}Tr\ln D^\dagger D$ of the fermion determinant. We examine them in turn.

6.1 Non-local regularizations

Regularizations in which the momenta of the quark propagators are limited in some way lead to non-local actions. They have several advantages. They usually regularize the whole theory and not only the fermion determinant or only its real part. They introduce extra terms in conserved currents, to which electroweak gauge fields couple (see section 9.4). The construction of conserved currents, in the presence of non-local regulators, then gives rise however to correct unregularized anomalies [19, 69, 70, 71].

Non-local regularizations however cause a serious problem. They lead to propagators which are only defined for euclidean (space-like) momenta. The analytic continuation, required to determine on-shell masses, remains an unsolved problem, except in the case of 3-momentum regularization, which breaks Lorentz covariance. Applications using non-local regularization can be found in [19, 20, 26, 35, 37, 38, 70, 72, 73, 74].

6.1.1 *Regularization by delocalized fields*

Consider the case where the interaction between the quarks has the form (4.12) expressed in terms of the *delocalized fields* (3.13):

$$V = -\int d_3r\left[\frac{g_s^2}{2}\left(\left(\overline{\psi}\tau_a\psi\right)^2+\left(\overline{\psi}i\gamma_5\tau_a\psi\right)^2\right)+\frac{g_v^2}{2}\left(\overline{\psi}\gamma_\mu\tau_a\psi\right)^2+\frac{g_a^2}{2}\left(\overline{\psi}\gamma_5\gamma_\mu\tau_a\psi\right)^2\right] \quad (6.1)$$

The euclidean 4-vector notation (4.21) is used. The delocalized field $\psi(x)$ is related to the quark field $q(x)$ by the equation (3.13). It can be viewed as a quark field with limited Fourier components $q(k)$:

$$\psi(x) = \sum_k \int dy \ \langle x|k\rangle r\left(k^2\right)\langle k|y\rangle q(y) \equiv \sum_k \langle x|k\rangle r\left(k^2\right) q(k) \tag{6.2}$$

where $\langle x|k\rangle$ are the plane waves (3.31). The regulator $r\left(k^2\right)$ is a function of the euclidean momentum $k^2 > 0$ and it decreases with k^2 at least as fast as $1/k^2$. It has a range characterized by a cut-off parameter Λ so that the sums over momenta are effectively limited to $k < \Lambda$. For instanton induced interactions [26], $r^2\left(k^2\right) \sim 1/k^6$ for large k, as displayed in equation (3.12). A study of the quark propagator obtained by solving the Schwinger–Dyson equation with effective gluon exchange interactions suggests that $r^2\left(k^2\right)$ decreases as $1/k^2$ with logarithmic corrections [20, 21, 22]. The limiting case of a sharp 4-momentum cut-off occurs when the regulator takes the form of a step function:

$$r\left(k^2\right) = \theta\left(\Lambda^2 - k^2\right) \tag{6.3}$$

The identities (4.13) and (4.14) can be applied to quartic forms of the delocalized ψ fields. We can express the right hand sides of these identities in terms of the original quark fields q by writing:

$$\int d_4x\bar{\psi}(x)\tau_a\psi(x)S_a(x) = \int d_4xd_4yd_4z\bar{q}(y)\langle y|r|x\rangle\tau_aS_a(x)\langle x|r|z\rangle q(z) \tag{6.4}$$

The euclidean action (4.23) acquires the regularized form:

$$I(S,P,V,A) = -Tr\ln\left(-i\partial_\mu\gamma_\mu + m + rSr + i\gamma_5rPr + \gamma_\mu rV_\mu r + \gamma_5\gamma_\mu rA_\mu r\right)$$

$$+\frac{1}{2g_s^2}\int d_4x\left(S_a^2(x) + P_a^2(x)\right) + \frac{1}{2g_v^2}\int d_4xV_{\mu a}^2(x) + \frac{1}{2g_a^2}\int d_4xA_{\mu a}^2(x) \tag{6.5}$$

where we used the shorthand notation $S \equiv S_a\tau_a$, $P \equiv P_a\tau_a$ etc. In other words, the interaction, which is expressed as a local interaction of the delocalized fields ψ with the scalar field S_a, is an interaction of the quark fields q with the non-local field rS_ar.

The form (6.5) of the action regularizes not only the action but the whole model. To see this, consider expanding the action about the translationally invariant point:

$$S = M \qquad P = 0 \qquad V = 0 \qquad A = 0 \tag{6.6}$$

It is written in the form:

$$I(S,P,V,A) = I(S=M,P=0,V=0,A=0)$$
$$-Tr\ln\left(1+\frac{r^2}{-i\partial_\mu\gamma_\mu+m+r^2M}\left(S-M+i\gamma_5P+\gamma_\mu V_\mu+\gamma_5\gamma_\mu V_\mu\right)\right)$$
$$+\frac{1}{2g_s^2}\int d_4x\left(S_a^2-M_a^2+P_a^2\right)+\frac{1}{2g_v^2}\int d_4xV_{\mu a}^2+\frac{1}{2g_a^2}\int d_4xA_{\mu a}^2 \tag{6.7}$$

When the form (6.7) is inserted into the expression (4.19) of the partition function, the latter can be expanded in powers of the coupling constants g^2. This yields an expansion of the partition function in terms of Feynman graphs (see section 8.2), composed of local interactions joined by regularized quark propagators g, which are diagonal in momentum space:

$$g\left|k\right\rangle = \frac{r^2\left(k^2\right)}{k_\mu\gamma_\mu+m+r^2\left(k^2\right)M}\left|k\right\rangle \tag{6.8}$$

If $r^2\left(k^2\right)$ decreases faster than $\frac{1}{k^2}$ for euclidean momenta, then all the Feynman graphs contributing to the partition function converge. The regulator leads to a model which is free of ultraviolet divergences. When the regulator has the form (6.3) of a step function, the Feynman graphs have propagators which vanish when $k^2<\Lambda^2$. Such prescriptions are used, for example, in [15, 55]. A similar argument will show that non-local interactions of the form (3.28), which lead to actions of the form (3.33), also regularize the theory.

A related but not equivalent regularization consists in using the delocalized fields (6.2) to express the full action (4.6) instead of only the quark–quark interaction (4.12). This regularization modifies the meaning of the trace which defines the quark loop, as follows:

$$-Tr\ln D = -tr\sum_k r^2\left(k^2\right)\left\langle k\left|D\right|k\right\rangle \tag{6.9}$$

where tr denotes a trace over the discrete quantum numbers of the quark fields. This again leads to Feynman graphs in which the euclidean 4-momenta are cut off by the regulator $r\left(k^2\right)$. The regularization (6.9) differs from the one discussed at the beginning of this section in that even the unperturbed action, obtained by setting $V=0$, is regularized.

6.1.2 *Quarks which cannot materialize on-shell*

The problem with non-local regularizations is that they are only defined for euclidean momenta, that is, for space-like momenta such that $k^2\equiv k_\mu k_\mu>0$. When, for example, we calculate *on-shell* meson masses, we search for poles of propagators which occur for time-like momenta $k^2<0$ (see section 7.2.1). This requires analytic continuation of the propagators, which can be fouled up by the regulators $r\left(k^2\right)$. The latter can introduce extra unphysical poles. The problem already arises at the level of the quark propagator.

Consider, for example, the quark propagator (6.8) calculated with an exponential regulator:

$$r^2\left(k^2\right) = e^{-\frac{k^2}{2\Lambda^2}} \tag{6.10}$$

For simplicity, we set the current quark mass $m = 0$. The quark propagator becomes:

$$g(k) = \frac{r^2\left(k^2\right)}{k_\mu\gamma_\mu + r^2\left(k^2\right)M} = e^{-\frac{k^2}{2\Lambda^2}} \frac{-k_\mu\gamma_\mu + Me^{-\frac{k^2}{2\Lambda^2}}}{k^2 + M^2 e^{-\frac{k^2}{\Lambda^2}}} \tag{6.11}$$

In the absence of the regulator or, equivalently, in the limit $\Lambda \to \infty$ of an infinite cut-off, the quark propagator would have a single pole occurring at $k^2 = -M^2$. This pole represents a free on-shell quark with a constituent quark mass M. When the cut-off is finite, the quark propagator has poles which occur for momenta k^2 such that $k^2 + M^2 e^{-\frac{k^2}{\Lambda^2}} = 0$. Let us scale the momenta to $t = k^2/\Lambda^2$ and the mass M to $\mu = M/\Lambda$. The poles of the quark propagator are given by the solutions of the equation:

$$\mu^2 e^{-t} = -t \tag{6.12}$$

When $\mu \equiv \frac{M}{\Lambda} < \frac{1}{\sqrt{e}} = 0.606$, the equation has two real roots. When $\mu = 0.606$ a single root appears at $t \equiv \frac{k^2}{\Lambda^2} = -1$. This limiting case is illustrated Fig. 6.1.

When $\mu \equiv \frac{M}{\Lambda} > 0.606$, no real roots occur and this suggests that quarks no longer materialize on-shell.

Propagators displaying such behavior have been studied phenomenologically [37, 38]. They also emerge as solutions of the Schwinger–Dyson equation with modified gluon propagators [74, 75, 76], and with quarks propagating in random colored background fields [77, 78]. Recently, a quark propagator free of poles on the real k^2 axis, of the form:

$$\int_{\frac{1}{\Lambda^2}}^{\frac{1}{\mu^2}} dse^{-s\left(k^2+M^2\right)} = \frac{1}{k^2+M^2}\left(e^{-\frac{k^2+M^2}{\Lambda^2}} - e^{-\frac{k^2+M^2}{\mu^2}}\right) \tag{6.13}$$

has been derived from a proper-time regularized action (6.18), in which both low and high energy cut-offs are introduced [79]. Quark propagators of purely exponential form have also been used [80].

It has been suggested that models in which quarks cannot materialize on-shell effectively describe confined quarks. The problem with such models is, however, that the poles of the quark propagators usually get shifted to complex values of

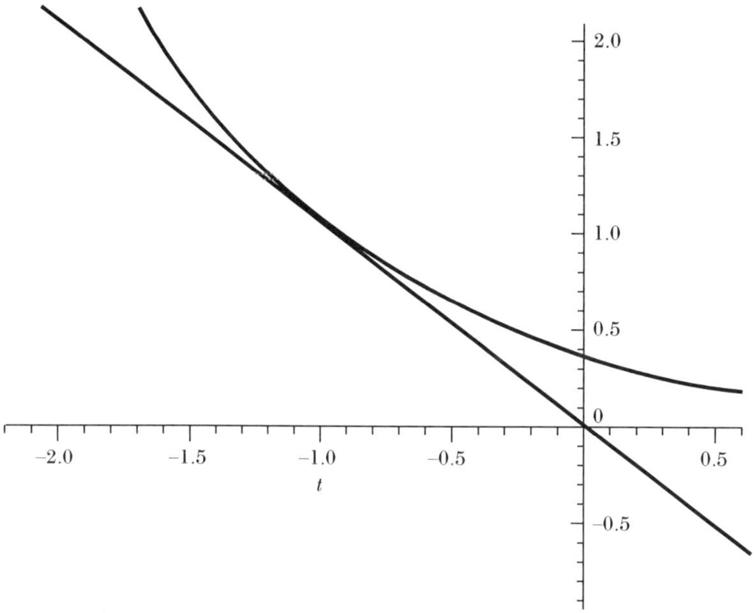

FIG. 6.1. Graphical solution of eqn (6.12) in the limiting case where $\mu = 1/\sqrt{e}$

t. This would herald an instability of the vacuum with respect to the addition of a single quark and, worse, the poles in the complex plane can foul up the meson propagators, when the latter are analytically continued to reach on-shell meson masses. It is however quite possible that, in an as yet unknown effective confining theory, the quark propagators will not have poles on the real negative k^2 axis and singularities elsewhere. We will then have to learn how to deal with such effective theories. Our ignorance as how to continue propagators in the complex k^2 plane reflects our ignorance of the confining mechanism [75].

6.2 3-momentum cut-offs

Problems related to analytic continuation of propagators can be sidestepped by cutting off the 3-momenta $|\boldsymbol{k}|$ of the quark propagators, rather than the 4-momenta $\sqrt{k_\mu k_\mu} = \sqrt{k_0^2 + \boldsymbol{k}^2}$. This can be achieved with non-local interactions by assuming, for example, that the regulators of the non-local interactions (3.28) or (3.35) depend only on the 3-momentum $\boldsymbol{k}$, as done, for example in [35]. Alternatively, the quark fields could be expanded on a basis of three-dimensional plane wave states which are limited to $|\boldsymbol{k}| < \Lambda$:

$$q(\boldsymbol{r}) = \sum_{|\boldsymbol{k}|<\Lambda} \langle \boldsymbol{r}|\boldsymbol{k}\rangle \, q_{\boldsymbol{k}} \tag{6.14}$$

This regularization is, in fact, a Hilbert space restriction. All states in which a quark has a momentum $\boldsymbol{k}$ exceeding Λ are eliminated. This is a mathematically very clean regularization. It does not lead to an instability of the vacuum and the propagators have only poles corresponding to on-shell particles. Unfortunately, when such a regularization is used, the Nambu Jona-Lasinio model fails to bind quarks so as to form solitons.

Regularizations with 3-momentum cut-offs also cause problems. They break Lorentz covariance. Thus, a meson, which is described as a $q\bar{q}$ excitation, cannot have a centre of mass momentum $|\boldsymbol{k}|$ exceeding 2Λ. This is unphysical and it leads to distortions of π-meson thermal distributions. When the temperature becomes comparable to the cut-off, the pressure exerted by the pions ceases to be proportional to T^4. This is why meson masses are usually calculated with zero momentum in the lab frame. A suggested way out of this difficulty, at least in the meson sector, consists in making the regulator $r(x-y)$ (3.28) or (3.35) depend only on the transverse component $x_\perp - y_\perp$ which is orthogonal to the center-of-mass 4-momentum of the interacting particles [36, 81]. This method has recently been applied to the calculation of the radial excitations of the pion [82].

It is simple to see why a 4-momentum cut-off does not limit the center-of-mass momentum of *on-shell* mesons. Indeed it yields meson propagators which are functions of q^2 and the poles occuring at $q^2 = -m^2$ yield the masses m of on-shell particles, with no restriction on their momentum $\vec{q}$ (see section 7.2.1).

6.3 Regularizations of the real part of the fermion determinant

Regularizations which regularize only the real part of the fermion determinant are frequently used. The reason for leaving the imaginary part unregularized is dictated by phenomenology. When the vector fields include electroweak fields, the imaginary part gives rise to anomalous processes, which are known to fit experiment if they are calculated with an infinite cut-off. This is further discussed in section 9.5.

Note however that, when non-local regularization is used for both the real and the imaginary part of the action (section 6.1.1), the anomalous processes remain unregularized [19, 69, 70, 71].

The trouble with regularizing only the real part of the fermion determinant is that the regularized action becomes a non-analytic and often singular functional of the fields. As a result, perturbation theory is no longer analytic in the coupling constants (see section 12.4). This can lead to problems when strongly interacting vector fields are included in soliton calculations (see section 12.10).

A comparison between various regularization schemes *of the real part of the action* can be found in [83]. (The reader should be warned that the 3- and 4-momentum regularizations discussed in [83] refer only to the regularization of the real part of the action. They are not to be confused with the regularizations discussed in section 6.1.1.)

6.3.1 *Proper-time regularization*

Proper-time regularization is the most commonly used regularization of the real part of the fermion determinant, essentially because it yields a finite local action of the fields and also because, applied to the Nambu Jona-Lasinio model, it allows for the existence of a soliton in the form of a bound state of quarks (chapter 12). Proper-time *renormalization* was introduced by Schwinger in order to maintain gauge invariance in perturbative quantum electrodynamics [84]. Proper-time *regularization* uses the same formalism but it differs from the corresponding renormalization in that it maintains a finite proper-time cut-off. It shares the property of maintaining gauge invariance of the fermion determinant. Although it leads to meson propagators with spurious unphysical poles [85], the latter occur off the real axis, so that proper-time regularized meson propagators do allow a calculation of on-shell meson masses.

Proper-time regularization only regularizes the quark loop term. When the proper-time regularized action is used to evaluate the partition function (4.19), only the classical (saddle point) partition function is finite. Ultraviolet divergences remain in higher order loop contributions [86, 87]. It has no microscopic justification. No derivation of a low energy effective theory is known to lead to a proper-time regularization.

Proper-time regularization consists in first separating the fermion determinant term of the euclidean action (4.23) into real and imaginary parts:

$$-Tr \ln D = -\frac{1}{2} Tr \ln D^\dagger D + \frac{1}{2} Tr \ln D^\dagger / D \tag{6.15}$$

where D is the Dirac operator (4.22). Only the real part is regularized and the imaginary part is left unregularized:

$$-Tr_\Lambda \ln D = \frac{1}{2} Tr \int_{\frac{1}{\Lambda^2}}^{\infty} \frac{ds}{s} e^{-sD^\dagger D} + \frac{1}{2} Tr \ln D^\dagger / D \tag{6.16}$$

The expression (6.16) reduces to the expression (6.15) in the limit $\Lambda \to \infty$, as can be seen from the expansion in powers of $\frac{1}{\Lambda^2}$:

$$\frac{1}{2} \int_{\frac{1}{\Lambda^2}}^{\infty} \frac{ds}{s} e^{-sD^\dagger D} = -\frac{\gamma}{2} - \frac{1}{2} \ln \frac{D^\dagger D}{\Lambda^2} + \frac{D^\dagger D}{2\Lambda^2} + \ldots \tag{6.17}$$

where $\gamma = 0.5772$ is Euler's constant. The variable s in (6.16) is called the *proper-time variable*. It has the dimension E^{-2} of an inverse energy squared. Ultraviolet divergences are eliminated by cutting off the low values of the proper-time s. In the regularization (6.16), a sharp cut-off $s < \frac{1}{\Lambda^2}$ is used. The sharp cut-off could, of course, be replaced by a smooth function of s. The proper-time regularized action (4.23) is thus:

$$I(S,P,V,A) = \frac{1}{2}Tr\int_{\frac{1}{\Lambda^2}}^{\infty}\frac{ds}{s}e^{-sD^\dagger D} + \frac{1}{2}Tr\ln D^\dagger/D$$

$$+\frac{1}{2g_s^2}\int d_4x\left(S_a^2+P_a^2\right)+\frac{1}{2g_v^2}\int d_4xV_{\mu a}^2+\frac{1}{2g_a^2}\int d_4xA_{\mu a}^2$$

$$D = -i\partial_\mu\gamma_\mu + m + S_a\tau_a + i\gamma_5P_a\tau_a + \gamma_\mu V_{\mu a}\tau_a + \gamma_5\gamma_\mu A_{\mu a}\tau_a \tag{6.18}$$

If we assume that the vector fields V and A transform as gauge fields, then the proper-time regularized fermion determinant is gauge-invariant. This has two consequences. First, the usual expressions $(\bar{q}\gamma_\mu q, ...)$ of conserved currents, as derived from local lagrangians, still apply. This is important if the vector fields include electroweak gauge fields for example. Second, the regularized fermion determinant makes no contribution to the mass terms to the vector fields. An example is given in section 9.3.1. The mass terms of vector fields, such as the last two terms of (6.18), break gauge invariance (strongly interacting vector mesons are not gauge bosons). Electroweak fields should not be included in the vector mass terms.

6.3.2 *Pauli–Villars regularization*

Pauli–Villars regularization of the real part of the action is another method borrowed from renormalization. It was originally proposed as a renormalization of the full fermion determinant (not just the real part) which preserves gauge invariance [88]. However, when it is applied to the full fermion determinant, it breaks chiral symmetry because it introduces massive fields. Nonetheless, when it is used to regularize only the real part of the fermion determinant, chiral symmetry can be maintained.

In Pauli–Villars regularization, counter-terms are added to the real part of the fermion determinant $-\frac{1}{2}Tr\ln D^\dagger D$. The regularized real part of the action is [89]:

$$-\frac{1}{2}Tr_\Lambda\ln D^\dagger D = -\frac{1}{2}Tr\ln D^\dagger D - \frac{c_1}{2}Tr\ln\left(D^\dagger D+\Lambda_1^2\right) - \frac{c_2}{2}Tr\ln\left(D^\dagger D+\Lambda_2^2\right) \tag{6.19}$$

where $\Lambda_1$and Λ_2 are cut-off parameters, with the dimension E of an energy, and where $c_1$and c_2 are dimensionless parameters, chosen such as to eliminate the quadratic and logarithmic divergences. To determine these coefficients, we can expand the action (6.19) about a point such as (5.7) and we choose the coefficients c_i so as to kill the divergences. It is a simple exercise to check that the quadratic and logarithmic divergences are cancelled when the coefficients c_1 and c_2 are equal to:

$$c_1 = -\frac{\Lambda_2^2}{\Lambda_2^2 - \Lambda_1^2} \qquad c_2 = \frac{\Lambda_1^2}{\Lambda_2^2 - \Lambda_1^2} \tag{6.20}$$

The fact the c_i coefficients have opposite sign means that one of the counter-terms in (6.19) represents a system of Dirac particles with boson statistics. This leads to an instability of the vacuum. Indeed, for such a system, energy is *gained* when $q\bar{q}$ excitations of energy $\sim 2\Lambda_i$ or larger are produced. When meson correlation functions are calculated, for example, these instabilities appear at *time-like* momenta ($k_\mu k_\mu < 0$) which exceed the cut-off. Therefore on-shell meson *masses* exceeding the cut-off Λ cannot be calculated, not a serious restriction.

However, the Pauli–Villars regularization offers the great advantage of being free from singularities in the complex k^2 plane, which occur in the case of proper-time regularization and of non-local regularizations. In spite of this advantage, surprisingly few applications have used the Pauli–Villars regularization [89, 90, 91].

7

CORRELATION FUNCTIONS: BASIC PROPERTIES

Correlation functions are ground state or thermal expectation values of time-ordered products of operators. Two-point correlation functions are also commonly called propagators or Green's functions. They allow us to study the response of a system to specific probes. The poles of correlation functions correspond to the eigenvalues of the hamiltonian and to on-shell masses of mesons and baryons. The residues at the poles correspond to transition matrix elements of operators between the ground state and the excited states. Correlation functions are thus closely related to experimental data. The application of correlation functions to hadronic physics is reviewed in [9, 92, 93]. Correlation functions are also the objects which are calculated (it is often said that they are 'measured') in QCD lattice calculations [10, 94].

We first review some properties of correlation functions. We then discuss how to calculate them in the framework of the effective action formalism, which leads to symmetry conserving approximations, discussed in chapter 8.

7.1 Spectral decomposition of correlation functions

Let H be the hamiltonian and $\vec{P}$ be the momentum operator of a system. They act in Fock space. We define the following euclidean momentum 4-vector:

$$P_\mu = P^\mu = \left(i\left(H - E_0\right), \boldsymbol{P}\right) \tag{7.1}$$

where E_0 is the ground state energy of the system. When the hamiltonian H commutes with the momentum operator $\boldsymbol{P}$, they have common eigenvectors $|n\rangle$:

$$H\,|n\rangle = E_n\,|n\rangle\,, \quad \boldsymbol{P}|n\rangle = \boldsymbol{k}_n\,|n\rangle\,, \quad P_\mu\,|n\rangle = k_{n\mu}\,|n\rangle\,, \quad k_{n\mu} = \left(i\left(E_n - E_0\right), \boldsymbol{k}_n\right) \tag{7.2}$$

Let φ_a be a set of operators. The operators φ_a may, but need not, be dynamical variables of the system, such as the fields for example. The operator P_μ is the space–time translation operator, so that we can write:

$$\varphi_a\left(x\right) \equiv \varphi_a\left(\tau, \boldsymbol{r}\right) = e^{-ix_\mu P_\mu}\varphi_a e^{ix_\mu P_\mu} \equiv e^{\tau H - i\boldsymbol{r}.\boldsymbol{P}}\varphi_a e^{-\tau H + i\boldsymbol{r}.\boldsymbol{P}} \tag{7.3}$$

where $x_\mu = x^\mu = (\tau, \boldsymbol{r})$ is the euclidean space–time 4-vector.

The correlation function of two operators φ_a and φ_b is the ground state expectation value:

$$\langle x_1 a | K | x_2 b \rangle = \langle 0 | T [\varphi_a (x_1) \varphi_b (x_2)] | 0 \rangle - \langle 0 | \varphi_a (x_1) | 0 \rangle \langle 0 | \varphi_b (x_2) | 0 \rangle \tag{7.4}$$

where T orders the operators in the order of decreasing time variables:

$$\begin{aligned} T [\varphi_a (\tau_1, \boldsymbol{r}_1) \varphi_b (\tau_2, \boldsymbol{r}_2)] = \theta (\tau_1 - \tau_2) \varphi_a (\tau_1, \boldsymbol{r}_1) \varphi_b (\tau_2, \boldsymbol{r}_2) \\ \pm \theta (\tau_2 - \tau_1) \varphi_b (\tau_2, \boldsymbol{r}_2) \varphi_a (\tau_1, \boldsymbol{r}_1) \end{aligned} \tag{7.5}$$

The + sign applies to (boson) operators which commute at different times and the − sign applies to (fermion) anticommuting operators.

Let us assume that the ground state $|0\rangle$ of the system is translationally invariant:

$$P_\mu |0\rangle = 0 \qquad H |0\rangle = E_0 |0\rangle \qquad \boldsymbol{P} |0\rangle = 0 \tag{7.6}$$

In this case, K is a function of $x_1 - x_2$ and it is diagonal in momentum space. Using the plane waves (3.31) we have:

$$\langle x_1 a | K | x_2 b \rangle = K_{ab} (x_1 - x_2) \qquad \langle k a | K | k' b \rangle = \delta_{kk'} K_{ab} (k)$$

$$K (q) = \int d_4 x e^{i q_\mu x_\mu} K (x) \qquad K (x) = \frac{1}{\beta \Omega} \sum_q e^{-i q_\mu x_\mu} K (q) \tag{7.7}$$

Using the completeness relation satisfied by the common eigenstates of H and $\boldsymbol{P}$ we obtain the following spectral decompositions:

$$K_{ab} (x) = \sum_{n \neq 0} \Big(\theta (\tau) \langle 0 | \varphi_a | n \rangle e^{-\tau \omega_n + i \boldsymbol{k}_n . \boldsymbol{r}} \langle n | \varphi_b | 0 \rangle \Big)$$

$$\pm \theta(-\tau) \langle 0 | \varphi_b | n \rangle e^{\tau \omega_n - i \boldsymbol{k}_n . \boldsymbol{r}} \langle n | \varphi_a | 0 \rangle$$

$$K_{ab} (q) = \Omega \sum_{n \neq 0} \left(\delta_{-\boldsymbol{q}, \boldsymbol{k}_n} \frac{\langle 0 | \varphi_a | n \rangle \langle n | \varphi_b | 0 \rangle}{\omega_n - i q_0} \pm \delta_{\boldsymbol{q}, \boldsymbol{k}_n} \frac{\langle 0 | \varphi_b | n \rangle \langle n | \varphi_a | 0 \rangle}{\omega_n + i q_0} \right) \tag{7.8}$$

where Ω is the volume of the system and where

$$\omega_n = E_n - E_0 \geq 0 \tag{7.9}$$

are the excitation energies of the system. The relation (7.8) shows that, in the complex q_0 plane, the correlation function $K_{ab} (q)$ is analytic everywhere except for poles at $q_0 = \pm i (E_n - E_0)$ which occur on the imaginary axis. If the hamiltonian has a continuous spectrum, the corresponding poles form a cut. The analytical properties of correlation functions are simply a statement about the spectrum of the hamiltonian: that it is bounded from below and that it has only real eigenvalues.

The spectral decomposition (7.8) gives the form of the correlation function which would be obtained from an exact calculation, starting from a hermitian hamiltonian which is bounded from below. However, such forms are not always obtained when correlation functions are calculated from an effective theory and (or) in some approximation. For example, when a soliton is calculated in the mean field approximation (see chapter 12), the ground state of the system is not translationally invariant and the correlation function is not a function of $\boldsymbol{r}_1 - \boldsymbol{r}_2$ although it may be a function of $\tau_1 - \tau_2$. Furthermore, the analytic properties of correlation functions can be altered by the regulators which are introduced to regularize the fermion loop for example (see chapter 6). Even in the renormalizable Gell-Mann Levy model, the correlation function of the scalar field develops a pole on the real axis in the complex q_0 plane, which is called a Landau ghost (see section 11.3.3). This is why a spectral decomposition of the form (7.8) can not be taken for granted in a model calculation. It can however help to give a physical interpretation of the poles and singularities which do occur.

7.2 The interpretation of poles in the complex q^2 plane

In this section, we use the spectral decomposition (7.8) to examine the meaning of the poles of correlation functions. Different cases may occur, which we examine in turn.

7.2.1 *Poles on the negative real axis: on-shell particles*

Imagine that the calculated correlation function of an operator (or field) φ turns out to have the form:

$$K(q) = \frac{1}{q^2 + m^2} \tag{7.10}$$

The calculated correlation function is a function of the euclidean $q^2 = q_\mu^2$, which has a pole on the negative real axis of the complex q^2 plane. It is the Lorenz invariance of a model (including its regularization) which ensures that K is a function of $q^2 = q_0^2 + \boldsymbol{q}^2$ and not an independent function of q_0 and $\boldsymbol{q}$. Let us show that the pole at $q^2 = -m^2$ represents an on-shell particle with mass m.

In the complex q_0 plane, the correlation function (7.10) has poles on the imaginary axis. Indeed, we can write:

$$K(q) = \frac{1}{q_0^2 + \boldsymbol{q}^2 + m^2} = \frac{1}{2\sqrt{\boldsymbol{q}^2 + m^2}} \left(\frac{1}{\sqrt{\boldsymbol{q}^2 + m^2} + iq_0} + \frac{1}{\sqrt{\boldsymbol{q}^2 + m^2} - iq_0} \right) \tag{7.11}$$

If we compare the form (7.11) to the spectral decomposition (7.8), we interpret the correlation function (7.10) as follows. The operator φ excites eigenstates $|\boldsymbol{q}\rangle$ of the hamiltonian, which have momentum $\boldsymbol{q}$ and energy $\omega_q = \sqrt{\boldsymbol{q}^2 + m^2}$. When the operator φ is a boson operator, the states $|\boldsymbol{q}\rangle$ can be identified as on-shell mesons of mass m and momentum $\boldsymbol{q}$. The residues of the poles allow us to make the identification:

$$\langle 0 |\varphi| \boldsymbol{k}\rangle \langle \boldsymbol{k} |\varphi| 0\rangle = \frac{1}{2\omega_k \Omega} \tag{7.12}$$

which is, of course, compatible with the usual quantification of a scalar field:

$$\varphi(\boldsymbol{r}) = \sum_{\boldsymbol{q}} \sqrt{\frac{1}{2\omega_q}} \left(a_{\boldsymbol{q}}^{\dagger} \langle \boldsymbol{q} \mid \boldsymbol{r}\rangle + \langle \boldsymbol{r} \mid \boldsymbol{q}\rangle a_{\boldsymbol{q}} \right) \qquad |\boldsymbol{q}\rangle = a_{\boldsymbol{q}}^{\dagger} |0\rangle \tag{7.13}$$

where $\langle \boldsymbol{r} \mid \boldsymbol{q}\rangle$ are the plane waves (3.31). Fields are usually normalized such that in the vicinity of a pole, occurring at $q^2 = -m^2$ for example, the correlation function has the form (7.10).

Poles which describe on-shell particles have $q^2 = -m^2 < 0$, that is, negative squared euclidean momenta. Note that the euclidean momentum $q^2 = -m^2$ corresponds to a Lorenz momentum 4-vector $q_L^{\mu} = \left(\sqrt{\boldsymbol{q}^2 + m^2}, \boldsymbol{q}\right)$ which is *time-like*, that is, such that $q_L^{\mu} q_{L\mu} = m^2 > 0$.

7.2.2 *Poles on the positive real axis: Landau ghosts*

Ideally, correlation functions should only have poles and cuts situated on the negative real axis of the complex q^2 plane. However, certain models and most regularization schemes lead to meson propagators which have further poles situated in the complex plane. We proceed to discuss the interpretation of such poles.

Imagine that a calculated correlation function has a pole on the positive real axis in the complex plane of q^2:

$$K(q) = \frac{1}{q^2 - m^2} = \frac{1}{q_0^2 + \boldsymbol{q}^2 - m^2} = \frac{1}{m^2 - q^2} = \frac{1}{m^2 - q_0^2 - \boldsymbol{q}^2} \tag{7.14}$$

An example is given in section 11.3.3, in connection with the renormalized Gell-Mann Lévy model. Such a pole is called a Landau ghost. It occurs in electrodynamics for the photon propagator at energies $q^2 \approx m_e^2 e^{\frac{3\pi}{\alpha}} \approx 10^{556}$ GeV [5]. In section 11.3, we show that it also occurs in the Gell-Mann Lévy model at energies $q^2 \approx M^2 e^{\frac{4\pi^2}{N_c g^2}} \approx 1$ GeV [95, 96, 97].

We distinguish two cases depending on whether $\boldsymbol{q}^2 > m^2$ or $\boldsymbol{q}^2 < m^2$. If $\boldsymbol{q}^2 < m^2$, we can write the propagator (7.14) in the form (7.11):

$$K(q) = -\frac{1}{2\omega_q (q_0 - \omega_q)} - \frac{1}{2\omega_q (q_0 + \omega_q)} \qquad \omega_q = \sqrt{m^2 - \boldsymbol{q}^2} \tag{7.15}$$

Comparing this expression to the spectral decomposition (7.8) we would say that the operator (or field) φ excites states with momenta $\boldsymbol{k}$, (with $|\boldsymbol{k}| < m$)

and with pure imaginary excitation energies equal to $\pm i\sqrt{m^2 - \boldsymbol{k}^2}$. The ground state of the system would be unstable with respect to perturbations caused by the operator φ. Indeed, the operator φ would create perturbations which would exponentially increase and decrease in real time, depending on the initial conditions.

If $\boldsymbol{q}^2 > m^2$, then we could write the propagator in the form:

$$K(q) = -\frac{1}{2\omega_q(\omega_q + iq_0)} - \frac{1}{2\omega_q(\omega_q - iq_0)} \qquad \omega_q = \sqrt{\boldsymbol{q}^2 - m^2} \tag{7.16}$$

Comparing to the spectral decomposition (7.8), we might conclude that the operator φ excites states (or particles) with excitation energies $\omega_k = \sqrt{\boldsymbol{k}^2 - m^2}$. Such states, if they existed, would presumably have velocities $v = \frac{|\boldsymbol{k}|}{m} > 1$ larger than the speed of light. That is why Landau ghosts are sometimes referred to as *tachyonic modes.* In actual applications (see section 11.3), the residue of the pole is negative, which would imply that $\langle 0|\varphi|\boldsymbol{k}\rangle\langle\boldsymbol{k}|\varphi|0\rangle < 0$. The Landau ghost can therefore not be interpreted as a physical state.

The occurence of poles on the positive real axis has a much simpler interpretation. In section 7.4.2 we show that the second order variation of the effective action has the quadratic form (7.29) involving the inverse propagator. When a propagator of the form (7.14) occurs, a term of the form $\frac{1}{2}\delta\varphi(q)\left(m^2 - q^2\right)\delta\varphi(q)$ contributes to the effective action. This means that the translationally invariant stationary point of the effective action does not describe a state with minimum energy.

7.2.3 *Poles in the complex q^2 plane*

Assume that a calculated correlation function has a pole in the complex q^2 plane:

$$K(q) = \frac{1}{q^2 + z} = \frac{1}{q_0^2 + \boldsymbol{q}^2 + z} = \frac{1}{\left(\sqrt{\boldsymbol{q}^2 + z} + iq_0\right)\left(\sqrt{\boldsymbol{q}^2 + z} - iq_0\right)} \tag{7.17}$$

where z is a complex number. We could then write an expansion analogous to (7.11) and, comparing to (7.8), we would conclude that the operator φ excites states with complex excitation energies. If the imaginary part of the energy is negative, the excited state decays exponentially in real time. If it is positive, it grows exponentially.

Poles in the complex q^2 plane occur when quarks interact with non-local fields [37, 74] or when regulators are introduced to regularize the effective action [85]. They do not occur when a 3-momentum cut-off is used to regularize the theory (see section 6.2). At present, it is not clear whether models which produce propagators with poles in the complex q^2 plane should be discarded or not. So far, such poles have mostly been ignored. In some models, poles of the quark propagators are shifted away from the negative real axis (see section 6.1.2). Such a case might reflect (but not explain) the effect of confining forces [37, 74, 75].

7.3 Dispersion relations

The spectral decomposition (7.8) shows that when the hamiltonian of the system is bounded from below and when it has only real eigenvalues, the correlation function has only poles and cuts on the negative real axis in the complex q^2 plane. In general, the correlation function will have the form (with possible subtractions):

$$K(q) = \int_0^\infty ds \frac{\rho(s)}{q^2 + s} \tag{7.18}$$

where $\rho(s)$ is a positive definite function called the *spectral density*. It represents the probability of exciting a state with center-of-mass energy s. In the particular case (7.10), we have $\rho(s) = \delta(s - m^2)$. Usually a threshold m_0^2 occurs, above which the spectral density produces a cut, that is, a continuum of poles. In this case, the correlation function will typically have the form:

$$K(q) = \frac{1}{q^2 + m^2} + \int_{m_0^2}^\infty ds \frac{\rho(s)}{q^2 + s} \tag{7.19}$$

The threshold m_0 indicates the energy above which new channels open, composed of on-shell particles and into which the meson can decay. Correlation functions of, say, vector operators have extra q-dependent factors occurring in the numerator, but the singularity structure is the same. For an extensive review of the use of correlation functions in hadronic physics, see [9].

All the experimental information about the spectrum of the excited states is contained in the spectral density function $\rho(s)$ and m. For correlation functions of vector currents $\bar{q}\gamma_\mu q$, for example, the spectral density $\rho(s)$ may be measured by observing the e^+e^- annihilation into hadrons. Ideally, a model should aim at calculating $\rho(s)$ and m, which can then be directly compared to experiment.

Unfortunately, this is not always possible. For example, lattice QCD calculations are only able to calculate correlation functions for euclidean momenta $q^2 > 0$, whereas the experimental data occur for $q^2 < 0$. Models which are based on the instanton structure of the vacuum [9, 26, 31, 92] are also limited to the calculation of correlation functions with momenta $q^2 > 0$. The same limitation applies to Lorenz invariant non-local regularizations (see chapter 6). (It does not apply to proper-time and Pauli–Villars regularizations which regulate only the real part of the fermion determinant.) Grossly speaking, the problem is that theorists are often only able to calculate correlation functions for $q^2 > 0$ while the experimental data lies at $q^2 < 0$. A model or a theory which does allow us to make an analytic continuation of correlation functions to $q^2 < 0$, where the singularities corresponding to on-shell particles lie, is hard put to explain how quarks bind so as to form on-shell hadrons.

Failing this, dispersion relations of the form (7.19) are used as follows. The measured spectral density $\rho(s)$ and the measured on-shell mass m are used as an input. The dispersion relation (7.19) is then used to deduce the 'experimental' correlation function with $q^2 > 0$. The latter is then compared to the calculated one. There is a certain art in performing this task. Various subtractions can be performed and Borel transforms can be used to emphasize those parts of the spectral density which are known best [94].

However, model calculations of correlation functions with $q^2 > 0$ can be directly compared with those which are obtained in lattice calculations. See [98] for example.

Of course, much information is lost when we only look at correlation functions with $q^2 > 0$. The spectral density has structure on the $q^2 < 0$ axis, indicating bound states of quarks, most of which is smeared out when viewed from the negative q^2 axis. This becomes quite obvious when one compares Figs 1 and 2 of [9] for example. It is very difficult to reconstruct $\rho(s)$ from the correlation function $K(q)$ on the $q^2 > 0$ axis. One can check this by using the Nambu Jona-Lasinio model to calculate the correlation function in the ρ meson channel for both the $q^2 > 0$ and $q^2 < 0$, to leading order in N_c. When the vector coupling constant g_v^2 is strong enough to allow for a bound state in the ρ channel, a peak appears on the $q^2 < 0$ axis followed by a continuum. The latter is a $q\bar{q}$ continuum (and not the observed 2-pion continuum). If the vector coupling constant g_v^2 is made weaker, so that the ρ is no longer bound, then the peak acquires a large width due to its decay into $q\bar{q}$ pairs (and not into 2 pions). However, when the correlation function is viewed on the $q^2 > 0$ axis, the disappearance of the bound state is barely visible. Even in the absence of a bound state, a reasonably good fit can be obtained by parametrizing the spectral density function in terms of a peak and a continuum, the peak being in reality absent [98, 99].

7.4 The generating functional and the effective action

We now consider the calculation of correlation functions within a model. Correlation functions of operators φ_a are conveniently calculated from their generating functional $W(j)$ defined thus:

$$e^{-W(j)} = Tre^{-\beta H}Te^{\int d_4x j_a(x)\varphi_a(x)} \tag{7.20}$$

where $\varphi_a(x)$ is defined in (7.3) and $\int d_4x = \int_0^\beta d\tau \int_\Omega d_3r$. The operator T orders the operators from left to right in decreasing times τ. We define the *expectation value of any operator* F:

$$\langle F\rangle = \frac{Tre^{-\beta H}F}{Tre^{-\beta H}} \underset{\beta\to\infty}{\to} \langle 0|F|0\rangle \tag{7.21}$$

where $\langle 0|F|0\rangle$ is the ground state expectation value. The linked correlation functions (7.4), also called propagators, are obtained as functional derivatives of the generating functional (7.20):

$$\langle\varphi_a(x)\rangle = -\left.\frac{\delta W(j)}{\delta j_a(x)}\right|_{j=0}$$

$$\begin{aligned}\langle x_1 a\,|K|\,x_2 b\rangle &= \langle T\left[\varphi_a(x_1)\,\varphi_b(x_2)\right]\rangle - \langle\varphi_a(x_1)\rangle\,\langle\varphi_b(x_2)\rangle \\ &= -\left.\frac{\delta^2 W(j)}{\delta j_a(x_1)\,\delta j_b(x_2)}\right|_{j=0}\end{aligned} \tag{7.22}$$

The ground state energy, including the contribution of the sources, is given by the expression:

$$E_0(j) = \frac{\partial W(j)}{\partial\beta} \tag{7.23}$$

7.4.1 *Definition of the effective action*

It is often more convenient to deal with expectation values $\langle\varphi_a(x)\rangle$ than with sources j. This may be achieved by constructing an *effective action* $\Gamma(\varphi)$ as follows:

$$\Gamma(\varphi) = W(j) + \int d_4 x j_a(x)\,\varphi_a(x) \tag{7.24}$$

The effective action $\Gamma(\varphi)$ becomes a functional of φ provided that we relate j and φ by the equation:

$$\varphi_a(x) = -\frac{\delta W(j)}{\delta j_a(x)} \tag{7.25}$$

In other words, we use (7.25) to express $j_a(x)$ in terms of $\varphi_a(x)$. When the equation (7.25) is satisfied, $\varphi_a(x)$ becomes equal to the expectation value $\langle\varphi_a(x)\rangle$. The effective action (7.24) may be viewed as a Legendre transform. What is known as the *effective potential* [100], is the effective action (7.24) calculated with constant fields. In section 8.3 we express the effective action (7.24) in terms of Feynman graphs which are irreducible with respect to cutting the interaction lines.

7.4.2 *Properties of the effective action*

The effective action (7.24) has properties similar to a classical action. Let us show that it is stationary with respect to variations of the fields φ. Consider a variation $\varphi \to \varphi + \delta\varphi$. The corresponding variation of the effective action is (in obvious abbreviated notation):

$$\frac{\delta\Gamma(\varphi)}{\delta\varphi} = \left(\frac{\delta W(j)}{\delta j} + \varphi\right)\frac{\delta j}{\delta\varphi} + j = j \tag{7.26}$$

where use was made of (7.25). Thus, when the system is in equilibrium, that is, when the sources j vanish, *the effective action* $\Gamma(\varphi)$ *is stationary with respect to variations of the fields* φ_a. The variational principle can also be stated as follows: *at the stationary point of the effective action, the values of the operators* $\varphi_a(x)$ *become equal to the expectation values* $\langle\varphi_a(x)\rangle$.

From (7.26) we see that the second order derivative of the effective action is:

$$\frac{\delta^2 \Gamma(\varphi)}{\delta\varphi\delta\varphi} = \frac{\delta j}{\delta\varphi} \tag{7.27}$$

We can then use (7.25), (7.22) and (7.27) to show that the propagator (7.22) is (in full notation):

$$\langle x_1 a |K| x_2 b\rangle = -\frac{\delta^2 W(j)}{\delta j_a(x_1)\,\delta j_b(x_2)} = \frac{\delta\varphi_b(x_2)}{\delta j_a(x_1)} = \left(\frac{\delta^2 \Gamma(\varphi)}{\delta\varphi_a(x_1)\,\delta\varphi_b(x_2)}\right)^{-1} \tag{7.28}$$

In other words, *the second order functional derivative of the effective action is equal to the inverse propagator of the operators* φ. This is a very useful property. Indeed, the search for poles of the propagator is equivalent to the search for the zeros of the inverse propagator. The latter is simply obtained by deriving twice the effective action with respect to φ. This is equivalent to a solution of the Bethe–Salpeter equation. When the correlation function is diagonal in momentum space, it acquires the form (7.7).

In practice, the inverse correlation function K^{-1} is calculated by expanding the effective action to second order in powers of the operators, around a stationary point φ. For a translationally invariant system, the expansion has the form:

$$\Gamma(\varphi + \delta\varphi) = \Gamma(\varphi) + \frac{1}{2\beta\Omega}\sum_q \delta\varphi_a(q)\, K_{ab}^{-1}(q)\,\delta\varphi_b(-q) + \ldots \tag{7.29}$$

where $K(q)$ is defined in (7.7) and $\delta\varphi(q) = \int d_4x e^{iq_\mu x_\mu}\delta\varphi(x)$. Poles of the correlation function, from which on-shell particle masses are deduced, correspond to the zero eigenvalues of the matrix $K_{ab}^{-1}(q)$.

7.4.3 *The effective action* $\Gamma(\varphi)$ *in the one-loop approximation*

In the following, we shall assume that the dynamical variables of the system are the fields $\varphi_a(x)$. We assume that the system is described by an action $I(\varphi)$, which could have the form (4.23), for example. The path integral for the generating functional (7.20) is then [41, 42]:

$$e^{-W(j)} = \int D(\varphi)\, e^{-I(\varphi)+\int d_4x j_a(x)\varphi_a(x)} \tag{7.30}$$

Let us evaluate the path integral using the saddle point method. At the stationary point $\varphi = \varphi_{class}$, we have:

$$\left.\frac{\delta I(\varphi)}{\delta\varphi_a(x)}\right|_{\varphi=\varphi_{class}} = j_a(x) \tag{7.31}$$

We expand the argument of the exponential in (7.30) to second order in $\varphi - \varphi_{class}$ and we perform the corresponding gaussian integral using (4.10). We obtain thus the generating functional in the one-loop approximation:

$$W(j) = I(\varphi_{class}) - j\varphi_{class} + \frac{1}{2} Tr \ln \left. \frac{\delta^2 I(\varphi)}{\delta\varphi\delta\varphi} \right|_{\varphi=\varphi_{class}} \tag{7.32}$$

From the expression (7.32), we can deduce the following expression for the effective action (7.24):

$$\Gamma(\varphi) = I(\varphi_{class}) - j\varphi_{class} + \frac{1}{2} Tr \ln \left(\frac{\delta^2 I(\varphi_{class})}{\delta\varphi_{class}\delta\varphi_{class}} \right) + j\varphi \tag{7.33}$$

Assume that the fluctuation $\varphi - \varphi_{class}$ is small, of the order of ε. In many applications $\varepsilon \approx 1/N_c$. Then, since $I(\varphi) - j\varphi$ is stationary at $\varphi = \varphi_{class}$, we can replace $I(\varphi_{class}) - j\varphi_{class}$ by $I(\varphi) - j\varphi$ to order ε^2. Now $\frac{1}{2} Tr \ln \left(\frac{\delta^2 I(\varphi_{class})}{\delta\varphi_{class}\delta\varphi_{class}} \right)$ is also of order ε. Therefore, to order ε^2, the effective action (7.33) can be replaced by [5]:

$$\Gamma(\varphi) = I(\varphi) + \frac{1}{2} Tr \ln \left(\frac{\delta^2 I(\varphi)}{\delta\varphi\delta\varphi} \right) \tag{7.34}$$

The expression (7.34) is the effective action *in the one-loop approximation.* In section 8.6, we show that the one-loop effective action (7.34) is obtained by summing the ring graphs. The first term is simply the classical action, which is usually of order N_c. The second term is the contribution of the quantal vibrations of the fields. It is of order $(N_c)^0$ (see section 5.3). The leading order (in N_c) approximation, which consists in neglecting the second term of (7.34), is often referred to as the *Hartree approximation*, or the *mean field approximation.*

The one-loop approximation (7.34) to the effective action is a well-defined functional of the fields φ only if the operator $\frac{\delta^2 I(\varphi)}{\delta\varphi\delta\varphi}$ is positive definite. This is usually the case at a stationary point of the classical action $I(\varphi)$, but it is not necessarily the case for other values of the fields φ, such as the stationary point of the effective action $\Gamma(\varphi)$, for example. A way out of this difficulty is discussed in section 8.6, which makes use of the fact that it is consistent to evaluate the second term of the effective action either at φ_{class} or at a neighboring value φ. In order to avoid a certain confusion which occasionally arises, it is important to realize that the field φ, appearing in the effective action (7.34) or (7.24), is *not* the integration variable with which the path integral (7.30) is performed. It has a different meaning. At a stationary point of the effective action $\Gamma(\varphi)$, the field φ is equal to the expectation value of the field operator φ in the ground state of the system (or in the system at equilibrium at finite temperature). For other values, it can only be interpreted as the expectation value of the field operator in the presence of the source j. The whole point is that the effective action expresses the result of performing the path integral, to any given approximation, without having to refer to the source.

7.5 The generating functional of quark bilinear forms

Expectation values of quark bilinear operators $\bar{q}\Gamma_a q$, such as those listed in Table 2.2 for example, can be calculated by introducing source terms $-\eta_a\bar{q}\Gamma_a q$ to the Dirac operator, which becomes :

$$D \to -i\partial_\mu\gamma_\mu + m + \Gamma_a\varphi_a - \Gamma_a\eta_a \tag{7.35}$$

The generating functional (7.30) becomes:

$$e^{-W(j,\eta)} = \int D(\varphi)\, e^{-I(\varphi)+\int d_4x j_a(x)\varphi_a(x)+\int d_4x\eta_a(x)(\bar{q}\Gamma_a q)_x} \tag{7.36}$$

Expectation values and connected correlation functions of the quark bilinear operators are:

$$\langle(\bar{q}\Gamma_a q)_x\rangle = -\frac{\delta W(j,\eta)}{\delta\eta_a(x)} \qquad \left\langle T\left[(\bar{q}\Gamma_a q)_x (\bar{q}\Gamma_b q)_y\right]\right\rangle_c = -\frac{\delta^2 W(j,\eta)}{\delta\eta_a(x)\,\delta\eta_b(y)}$$

$$\left\langle T\left[\varphi_a(x)(\bar{q}\Gamma_b q)_y\right]\right\rangle_c = -\frac{\delta^2 W(j,\eta)}{\delta\eta_b(y)\,\delta j_a(x)} \tag{7.37}$$

Expectation values of quark bilinear operators can be expressed in terms of Feynman graphs (see section 8.2.3). They can also be calculated from the effective action (7.24), which is now:

$$\Gamma(\varphi,\eta) = W(j,\eta) + \int d_4x j_a(x)\,\varphi_a(x) \tag{7.38}$$

The expectation value of the quark bilinear form $(\bar{q}\Gamma_a q)_x$ is:

$$\langle(\bar{q}\Gamma_a q)_x\rangle = \frac{\delta\Gamma(\varphi,\eta)}{\delta\eta_a(x)} \tag{7.39}$$

In the one-loop approximation (7.34), the expectation value (7.39) becomes:

$$\langle(\bar{q}\Gamma_a q)_x\rangle = \frac{\delta I(\varphi,\eta)}{\delta\eta_a(x)} + \frac{1}{2}\frac{\delta^3 I(\varphi,\eta)}{\delta\eta_a(x)\,\delta\varphi_b(y)\,\delta\varphi_c(z)}\langle yb|K|zc\rangle \tag{7.40}$$

Summation and integration over repeated indices are assumed.The mixed correlation function $\left\langle T\left[\varphi_a(x)(\bar{q}\Gamma_b q)_y\right]\right\rangle_c$ can be derived from the following argument. The value of the field $\varphi_a(x)$ which makes the effective action stationary is equal to the ground state expectation value $\langle\varphi_a(x)\rangle = -\frac{\delta W(j,\eta)}{\delta j_a(x)}$. Therefore $\langle\varphi_a(x)\rangle$ is the solution of the equation:

$$\frac{\delta\Gamma(\varphi,\eta)}{\delta\varphi_a(x)} = 0 \tag{7.41}$$

Let φ be the solution of (7.41) for $\eta = 0$ and let $\varphi + \delta\varphi$ be the solution for a small η. We can then expand the equation (7.41) to first order in η:

$$\delta\varphi_a(x) = -\langle xa|K|yb\rangle \frac{\delta^2\Gamma(\varphi,\eta)}{\delta\varphi_b(y)\,\delta\eta_c(z)}\delta\eta_c(z) \tag{7.42}$$

It follows that the mixed correlation function is:

$$\langle xa|S|yb\rangle \equiv \left\langle T\left[\varphi_a(x)\left(\bar{q}\Gamma_b q\right)_y\right]\right\rangle_c = \frac{\delta^2\varphi_a(x)}{\delta\eta_b(y)} = -\langle xa|K|zc\rangle \frac{\delta^2\Gamma(\varphi,\eta)}{\delta\varphi_c(z)\,\delta\eta_b(y)} \tag{7.43}$$

from which one can easily deduce the one-loop approximation. In a translationally invariant system, we have $\langle xa|S|yb\rangle = S_{ab}(x-y)$ and the spectral decomposition of the mixed correlation function, analogous to (7.8), is:

$$S_{ab}(q) = \Omega\sum_{n\neq 0}\left(\delta_{-\boldsymbol{q},\boldsymbol{k}_n}\frac{\langle 0|\varphi_a|n\rangle\langle n|(\bar{q}\Gamma_b q)|0\rangle}{\omega_n - iq_0} \pm \delta_{\boldsymbol{q},\boldsymbol{k}_n}\frac{\langle 0|(\bar{q}\Gamma_b q)|n\rangle\langle n|\varphi_a|0\rangle}{\omega_n + iq_0}\right) \tag{7.44}$$

From (7.43) we deduce that:

$$S_{ab}(q) = K_{ac}(q)\,T_{cb}(q) \qquad \langle zc|T|yb\rangle \equiv \frac{\delta^2\Gamma(\varphi,\eta)}{\delta\varphi_c(z)\,\delta\eta_b(y)} \tag{7.45}$$

We can normalize the fields φ such that, in the vicinity of a pole $q^2 = -m^2$, the equation (7.10) holds, and $\langle 0|\varphi_a|n\rangle = \frac{1}{\sqrt{2\omega_q}}$. Then, if the correlation functions are diagonal in the indices a and b, the spectral decomposition (7.44) yields the following transition matrix element:

$$\langle n|(\bar{q}\Gamma_a q)|0\rangle = \sqrt{2\omega_q}T_{(aa)}\left(q^2 = -m^2\right) \qquad \omega_q = \sqrt{\boldsymbol{q}^2 + M^2} \tag{7.46}$$

where $|n\rangle$ is the state representing an on-shell particle with momentum $\boldsymbol{q}$. Such expressions can be used to calculate the π and K decay constants (9.29), for example, as well as the next to leading order contributions to the weak decay constants [87].

8

SYMMETRY CONSERVING APPROXIMATIONS

The one-loop approximation (7.34) of the effective action is but one of several symmetry conserving approximations. Other forms of symmetry conserving approximations exist which are briefly reviewed in this chapter. They were developed in the 1960s [101, 102]. For a review see [42]. Symmetry conserving approximations are conveniently be expressed in terms of Feynman graphs which have specific irreducibility properties. They proceed by expressing the partition function as a functional, either of the fields as in the effective action formalism, or of the quark propagators as explained in section 8.7. The self-energy of the quark propagator is then defined by the stationary point of the functional. When a symmetry of the action (typically chiral symmetry) is spontaneously broken by the self-energy, the correlation function acquires a pole at $q^2 = 0$ which corresponds to a zero mass Goldstone boson. Symmetry conserving approximations ensure that the existence of the Goldstone boson is maintained at any level of approximation.

8.1 A generic form of the action

Feynman graphs are most easily visualized if we use the following generic form of the euclidean action:

$$I(\varphi, \eta) = -Tr \ln (-i\partial_\mu \gamma_\mu + m + \Gamma_a \varphi_a - \Gamma_a \eta_a) - \frac{1}{2} \varphi V^{-1} \varphi - j\varphi \tag{8.1}$$

where:

$$\frac{1}{2} \varphi V^{-1} \varphi \equiv \frac{1}{2} \int d_4x d_4y \varphi_a(x) \left\langle xa \left| V^{-1} \right| yb \right\rangle \varphi_b(y) \tag{8.2}$$

The fields φ_a could be the fields $S_a, P_a, V_{\mu a}$ and $A_{\mu a}$ of the Nambu Jona-Lasinio action (4.23) for example. In this case the interaction V would have the form:

$$\langle xa |V| yb \rangle = -g_a^2 \delta_{ab} \delta(x - y) =$$

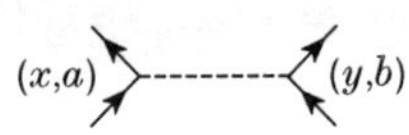

(8.3)

In the case of the Gell-Mann Lévy model, the interaction V could represent an unperturbed meson propagator. The interaction V is proportional to $1/N_c$. We have introduced source terms j and η which will be used to generate correlation functions.

8.2 Feynman graphs

The reader who wishes to familiarize himself with Feynman graph representation of generating functionals is referred to chapters 5 and 12 of [42]. We shall only briefly sketch the main argument. Unless stated otherwise, we restrict the discussion to Feynman graphs obtained for the unregulated theory. The method discussed below allows the reader to develop, for example, Feynman graph rules for the real part of the action with proper-time regularization. The graph analysis which leads to symmetry conserving approximations applies equally well to such cases.

8.2.1 *Unlabeled Feynman graphs for the generating functional*

We write the action (8.1) in the form:

$$I(\varphi,\eta) = -W_0(\eta) - \frac{1}{2}\varphi V^{-1}\varphi - Tr\ln\left(1+g\left(\Gamma_a\varphi_a\right)\right) - j\varphi \tag{8.4}$$

where:

$$W_0(\eta) = -Tr\ln\left(-i\partial_\mu\gamma_\mu + m - \Gamma_a\eta_a\right) \qquad g = \frac{1}{-i\partial_\mu\gamma_\mu + m - \Gamma_a\eta_a} \tag{8.5}$$

If the action is regularized in terms of delocalized fields, as in (6.5) for example, the quark propagator g acquires the form (6.8) and all the Feynman graphs become finite.

We insert the form (8.4) of the action into the path integral (7.30) and we expand the exponential in powers of $\Gamma_a\varphi_a$ and j:

$$e^{-W(j,\eta)} = e^{-W_0}\int D(\varphi)\, e^{\frac{1}{2}\varphi V^{-1}\varphi} e^{Tr\ln(1+g\Gamma_a\varphi_a)+j\varphi}$$

$$= e^{-W_0}\int D(\varphi)\, e^{\frac{1}{2}\varphi V^{-1}\varphi}\left(1+j\varphi+\frac{1}{2}(j\varphi)^2+...\right)\left(Trg\Gamma_a\varphi_a+\frac{1}{2}\left(Trg\Gamma_a\varphi_a\right)^2\right.$$
$$\left. - \frac{1}{2}Trg\Gamma_a\varphi_a g\Gamma_a\varphi_a+...\right) \tag{8.6}$$

Each term of the expansion (8.6) can be represented by a graph of the form:

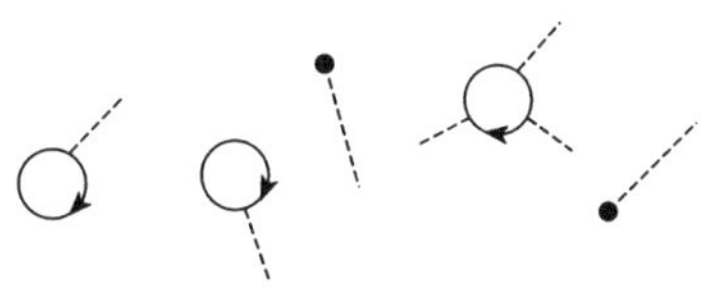

(8.7)

The graph consists of disconnected closed loops formed by oriented lines, called quark propagators. Each oriented line represents a quark propagator g. Between each quark propagator, there is a open dashed insertion line which represents an interaction $\Gamma_a\varphi_a$. Each closed loop contributes a factor equal to

$Trg\Gamma_a\varphi_a g\Gamma_a\varphi_a$.... The graphs consist also of open dashed lines with black dots, each of which represents a factor $j\varphi$.

It is essential that the exponential in the path integral (8.6) should be quadratic in the fields φ. This allows us to integrate the fields φ using the gaussian integration for which Wick's theorem applies (see appendix C). The gaussian integrals are of the form (4.10). Gaussian integrals of the form $\int D(\varphi)\,\varphi\varphi\varphi\varphi...$ can all be expressed in terms of the contraction:

$$\int D(\varphi)\,\varphi(x)\,\varphi(y)\,e^{\frac{1}{2}\varphi V^{-1}\varphi} = \langle x\,|V|\,y\rangle \int D(\varphi)\,e^{\frac{1}{2}\varphi V^{-1}\varphi} \tag{8.8}$$

Wherever no ambiguity can arise, we shall use the abbreviated notation:

$$x \equiv (x, a) \tag{8.9}$$

in order to make the expressions more legible. The result of the φ integration can be represented by graphs in which the open dashed lines join up pairwise in every possible way, so as to form dashed interaction lines, each one representing an interaction V. The interaction lines connect oriented quark propagators as well as source terms j. The disconnected graphs exponentiate into connected graphs in the usual way. The result is an expansion of the generating functional $W(j,\eta)$ in terms of *unlabeled connected graphs*:

$W(j,\eta) = W_0(\eta) +$ 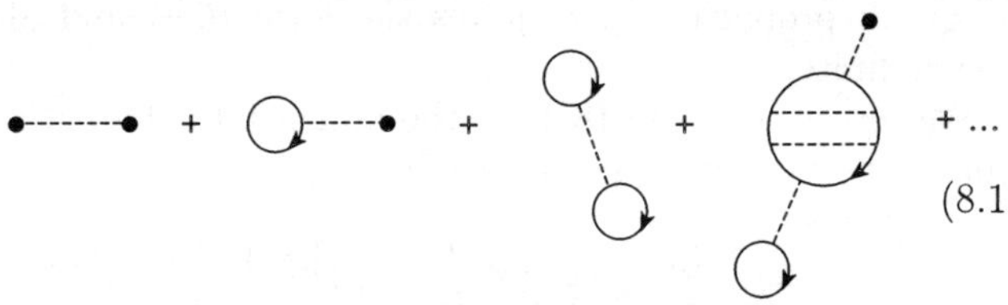 (8.10)

The generating functional is equal to the sum of all connected topologically distinct unlabeled graphs. Graphs are considered to be topologically distinct if they cannot be transformed into each other by translating the (dashed) interaction lines and/or by exchanging the extremities of the (dashed) interaction lines.

The rules for calculating the contribution of a given unlabeled connected graph are the following:

1. Assign labels $x, y, ...$ to the vertices, which are the extremities of the interaction lines.
2. Each (dashed) interaction line, joining vertices labeled x and y, contributes a factor $\langle x\,|V|\,y\rangle = \langle y\,|V|\,x\rangle$.
3. Each oriented (quark propagator) line, going from a vertex labeled x to a vertex labeled y, contributes a factor $\langle x\,|g|\,y\rangle = \left\langle x \left| \frac{1}{-i\partial_\mu\gamma_\mu + m - \Gamma_a\eta_a} \right| y \right\rangle$.
4. Each black dot at the extremity x of an interaction line contributes a factor $j(x)$.
5. Multiply the product of these factors by $(-)^{1+n_l+n_V+n_j}$ where n_l is the number of closed loops formed by the quark propagators, n_V is the number

of interaction lines, and n_j is the number of black dots at the extremities of open interaction lines.

6. Integrate and sum the continuous and discrete vertex labels $x, y, \ldots$
7. Divide the contribution by the symmetry factor S of the graph. The symmetry factor S is equal to the number of permutations of the vertex labels x and of the extremities of the interaction vertices which transform the graph into an identical graph. The determination of the symmetry factor occasionally requires some thought.

A graph, together with its contribution, is given below:

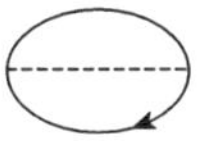

$$= \frac{1}{2} (-)^{1+1+1} \int d_4 x d_4 y \, \langle x |g| y \rangle \langle y |g| x \rangle \langle x |V| y \rangle \tag{8.11}$$

8.2.2 *Labeled Feynman graphs for correlation functions of the fields*

From (7.22) we deduce the following expansions for the expectation values of the fields and for the correlation functions in terms of labeled connected graphs:

$$\langle \varphi (x) \rangle = - \left. \frac{\delta W (j, \eta)}{\delta j (x)} \right|_{j=0} =$$

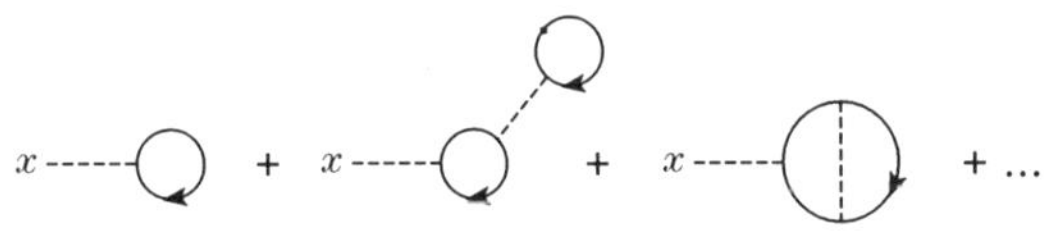

$$\langle x |K| y \rangle = \langle 0 |T [\varphi (x_1) \varphi (x_2)]| 0 \rangle_c = - \left. \frac{\delta^2 W (j, \eta)}{\delta j (x) \delta j (y)} \right|_{j=0}$$

$$- \langle x |V| y \rangle +$$

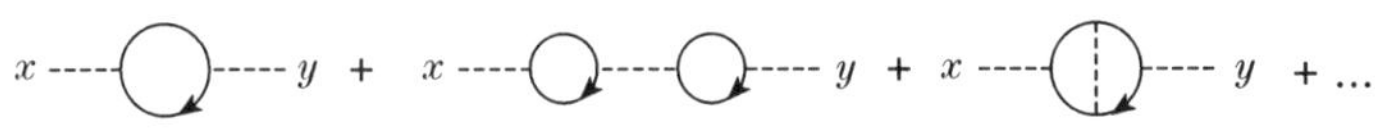

(8.12)

where $\langle \rangle_c$ means the connected expectation value defined in (7.22). The labeled vertices are *not* to be taken into account for the evaluation of the symmetry factor.

8.2.3 *Labeled Feynman graphs for correlation functions of quark bilinear operators*

The expectation values and correlation functions of quark bilinear forms, such as those listed in Table 2.2, can be obtained as functional derivatives of $W (j, \eta)$ with respect to the source η. A graph depends on η through the propagator g:

$$\frac{\delta g}{\delta \eta_a(x)} = g\,|x\rangle\,\Gamma_a\,\langle x|\,g \tag{8.13}$$

Thus a functional derivative with respect to $\eta_a(x)$ can be represented by a black dot labeled (x,a) inserted on an quark propagator g. We deduce the following expansion of the quark bilinear forms, in terms of labeled connected graphs:

$$\langle (\bar{q}\Gamma_a q)_x \rangle = - \left.\frac{\delta W(j,\eta)}{\delta \eta_a(x)}\right|_{\eta=0} =$$

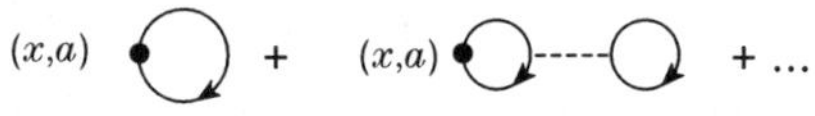

$$\langle xa\,|L|\,yb\rangle = \left\langle 0 \left| T\left[(\bar{q}\Gamma_a q)_x (\bar{q}\Gamma_b q)_y\right]\right| 0 \right\rangle_c = - \left.\frac{\delta^2 W(j,\eta)}{\delta \eta_a(x)\,\delta \eta_b(y)}\right|_{j=0}$$

$$=$$

(8.14)

The first terms on the right hand sides of (8.14) are due to the functional derivatives of $W_0(\eta)$.

If we compare the expansions (8.12) and (8.14) we can deduce the relations:

$$\langle \varphi_a(x) \rangle = - \int d_4 y\, \langle xa\,|V|\,yb\rangle\, \langle (\bar{q}\Gamma_a q)_x \rangle$$

$$K = -V - VLV \tag{8.15}$$

The correlation functions K and L have the same poles, so that either can be used to calculate on-shell masses for example. Graph expansions of mixed correlation functions $\left\langle 0 \left| T\left[\varphi_a(x)(\bar{q}\Gamma_b q)_y\right]\right| 0 \right\rangle_c$ can also easily be derived from the functional derivative given in equation (7.37). They can be used, for example, to calculate the pion decay constant f_π.

8.3 The Feynman graph expansion of the effective action

Gap equations and correlation functions which are next to leading order in N_c have recently been evaluated in a symmetry conserving approximation [86, 87], which is based on the one-loop approximation of the effective action (7.34), discussed in section 7.4. Symmetry conserving approximations based on approximate evaluations of the effective action (7.24) can be defined in terms of 1-interaction-line irreducible Feynman graphs. The reader who wishes to familiarize himself with the graph analysis and the graph resummations discussed in this section is referred to section 16.3 of [42].

Let $\Phi(\varphi)$ be the sum of 1-interaction-line irreducible graphs, which cannot be separated into two parts by cutting one (dashed) interaction line:

$\Phi(\varphi) =$  (8.16)

The graphs (8.16) are to be calculated with the rules given in section 8.2.1, with the exception that the quark propagators, represented by the oriented lines, contribute each a factor $D^{-1} = \frac{1}{-i\partial_\mu\gamma_\mu + m - \Gamma_a\eta_a + \Gamma_a\varphi_a}$ instead of $g = \frac{1}{-i\partial_\mu\gamma_\mu + m - \Gamma_a\eta_a}$. We say that the graphs (8.16) are calculated with *dressed quark propagators.* In the effective action formalism, the quark propagators are dressed with the local fields φ .

The effective action (7.24) can be expressed in terms of the 1-interaction-line irreducible graphs $\Phi(\varphi)$ as follows:

$$\Gamma(\varphi) = I(\varphi) + \Phi(\varphi) = -Tr \ln D - \frac{1}{2}\varphi V^{-1}\varphi + \Phi(\varphi) \qquad (8.17)$$

In spite of the fact that the irreducible graphs $\Phi(\varphi)$ are calculated with dressed quark propagators, a topological analysis of the graphs shows that the expression (8.17) includes the contribution of every unlabeled graph, calculated with undressed propagators g, exactly once [42].

The values of the fields φ, at a stationary point of the effective action (8.17), are equal to the expectation values of the fields in the ground state. Indeed, at a stationary point of the effective action (8.17), we have:

$$\varphi(x) = \int d_4y \langle x|V|y\rangle \frac{\delta}{\delta\varphi(y)} (-Tr \ln D + \Phi(\varphi))$$

$=$ 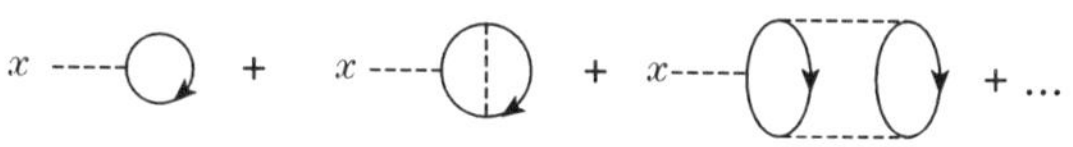

(8.18)

These are precisely the graphs (8.12) which contribute to $\langle\varphi(x)\rangle$. However, each graph in (8.18) includes all the 1-interaction-line reducible graphs (8.12) which are generated by making any number of local insertions to the quark propagators. The gap equation (5.14) becomes equivalent to the equation (8.18) when we set $\Phi(\varphi) = 0$. In this case, all the 1-interaction-line reducible graphs are neglected. The equation (8.18) may be viewed as a *generalized gap equation.*

The second order derivative of the effective action (8.17) is equal to the inverse correlation function K defined in (8.12):

$$K^{-1} = \frac{\delta\Gamma(\varphi)}{\delta\varphi\delta\varphi} = -V^{-1} - L_0 - \Pi \tag{8.19}$$

where:

$\langle x|L_0|y\rangle =$ 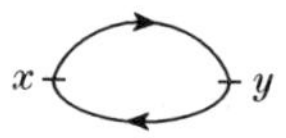

$= -\dfrac{\delta^2}{\delta\varphi(x)\,\delta\varphi(y)} Tr \ln D$

$\langle x|\Pi|y\rangle =$ 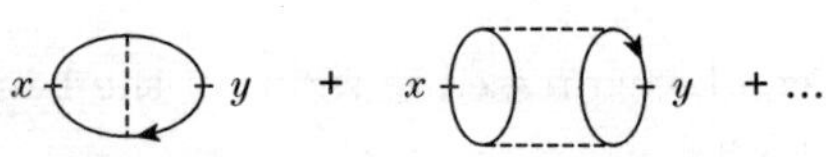

(8.20)

The term L_0 arises from the second order derivative of $-Tr \ln D$. From (8.19) we deduce the equation:

$$K = -V - V(L_0 + \Pi)K \tag{8.21}$$

$=$ 

(8.22)

These are precisely the graphs (8.12) which contribute to $\langle x|K|y\rangle = \langle 0|T[\varphi(x_1)\varphi(x_2)]|0\rangle_c$. From (8.15) we see that the integral equation for L, defined in (8.14) is:

$$L^{-1} = -(L_0 + \Pi)^{-1} - V \qquad L = -(L_0 + \Pi) - (L_0 + \Pi)VL \tag{8.23}$$

The analysis above acquires a practical value once it is realized that the properties above remain valid when $\Phi(\varphi)$ is calculated with *any subset* of 1-interaction-line irreducible graphs (8.16). Each subset defines a symmetry conserving approximation to the effective action (see section 8.4). One may even choose not to include any of the 1-interaction-line irreducible graphs. In that case, the effective action (8.17) reduces to the classical action and one is simply calculating the gap equation and the correlation functions to leading order in N_c, as in sections 5.2 and 8.5. If one chooses to include the ring graphs (8.33) in Φ, then the effective action (8.17) becomes identical to the effective action (7.34) calculated in the one-loop approximation (section 8.6). Once a given subset of graphs is chosen, it is essential to calculate both the gap equation and the correlation functions self-consistently, by performing the first and second order derivatives of the effective action (8.17). This ensures that the approximation is symmetry conserving, as shown in the following section.

8.4 The symmetry conserving property

We now show that when a continuous symmetry is spontaneously broken at a translationally invariant stationary point, the field correlation function develops a pole at $q^2 = 0$, which represents a zero mass *Goldstone boson.* This is typical of a symmetry conserving approximation.

Consider a continuous transformation of the fields:

$$\varphi_a(x) \to \bar{\varphi}_a(x) = \varphi_a(x) + \varepsilon T_{ab}\varphi_b(x) + ... \equiv \varphi_a(x) + \delta\varphi_a(x) \tag{8.24}$$

where T_{ab} is the generator of the transformation and ε an amplitude. The transformation could be a chiral transformation (4.30) for example. Assume that the transformation (8.24) leaves the action $I(\varphi)$ invariant. Then $I(\bar{\varphi}) = I(\varphi)$ to all orders in ε. The transformation (8.24) is said to express a symmetry of the action. If the action $I(\varphi)$ is invariant under the transformation (8.24), then, for example, the effective action (7.34) in the one-loop approximation is also invariant. Indeed, if $I(\varphi) = I(\bar{\varphi})$ for all φ, then $\frac{\delta^2 I(\varphi)}{\delta\varphi\delta\varphi} = \frac{\delta^2 I(\bar{\varphi})}{\delta\bar{\varphi}\delta\bar{\varphi}}$. Therefore:

$$\Gamma(\varphi) = \Gamma(\bar{\varphi}) \tag{8.25}$$

The contribution of any 1-interaction-line irreducible graph (8.16) is invariant, so that (8.25) is valid for the effective action calculated with any subset of irreducible graphs.

If the transformation is continuous, we can consider the expansion:

$$\Gamma(\bar{\varphi}) = \Gamma(\varphi) + \int d_4x\, \delta\varphi_a(x) \frac{\delta\Gamma(\varphi)}{\delta\varphi_a(x)} + \frac{1}{2}\int d_4x d_4y\, \delta\varphi_a(x) \left\langle xa \left| K^{-1} \right| yb \right\rangle \delta\varphi_b(y) + ... \tag{8.26}$$

where we used the definition (7.28) of the correlation function K. Assume that φ is a stationary point of the action. In this case the second term of (8.26) vanishes and the symmetry property (8.25) implies that the third term also vanishes. When the stationary point is translationally invariant, the inverse propagator has the form (7.7). The vanishing of the third term can then be written in the form:

$$\frac{1}{2\beta\Omega}\sum_q \varphi_c(q)\, \widetilde{T}_{ca} K^{-1}_{ab}(q)\, T_{bd}\varphi_d(q) = 0 \tag{8.27}$$

where $\varphi(q) = \int d_4 x e^{iqx} \varphi(x)$ and where we used the plane waves (3.31). If the stationary point φ is translationally invariant, then $\varphi_c(q)$ vanishes for all but $q = 0$. It follows that the matrix $K_{ab}^{-1}(q)$ has a zero eigenvalue at $q = 0$, belonging to the eigenvector $T\varphi$. This means that the propagator K has a pole at $q = 0$. From the analysis in section 7.2.1, the pole corresponds to a zero mass particle, which is called a Goldstone boson. Note that the Goldstone boson occurs only if the stationary point of the effective action *breaks the symmetry.* Indeed, if it did not, we would have $\delta\varphi = 0$ and the equation (8.27) would no longer imply the existence of a zero eigenvalue of K^{-1}.

The Goldstone boson which corresponds to spontaneously broken chiral symmetry is the pion. Its finite mass ($m_\pi = 138$ MeV) is due to the small but finite masses of the u and d quarks (see Table 4.2), which explicitly break the invariance of the action under the symmetry transformation.

When a soliton is calculated, the stationary point φ is no longer translationally invariant. In this case, the same analysis may be applied and it leads to the equation:

$$\frac{1}{2}\int d_4 x d_4 y \varphi_c(x) \widetilde{T}_{ca} \left\langle xa \left| K^{-1} \right| yb \right\rangle T_{bd} \varphi_d(y) = 0 \tag{8.28}$$

Since $\varphi_c(x)$ is not translationally invariant, we can only conclude from equation (8.28) that the inverse propagator K^{-1} has a zero eigenvalue. Soliton solutions are usually time-independent, in which case $\varphi_c(x) = \varphi_c(\boldsymbol{r})$. This means that a state of zero energy exists, the wavefunction of which is $T_{bd}\varphi_d(y)$. Such a state is called a *spurious state* (see section 12.7). Spurious states are typically encountered in the random phase approximation of excited states of nuclei [42, 103, 104].

8.5 The leading order contribution in N_c

We first consider the case where we neglect the second term of the effective action (7.34). The effective action reduces then to the classical action (8.1), which is the leading order term in N_c . This is equivalent to neglecting all the 1-interaction-line irreducible graphs (8.16).

The gap equation (8.18), obtained by setting to zero the first order derivative of the action $I(\varphi)$ with respect to the field $\varphi_a(x)$, can be represented graphically as follows:

$$\varphi_a(x) = -\int d_4 y \left\langle xa \left| V \right| yb \right\rangle tr \left\langle x \left| \frac{1}{D}\Gamma_b \right| x \right\rangle =$$ 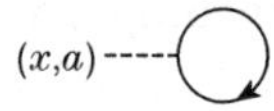

(8.29)

where tr is a trace over the discrete color, flavor and Dirac indices of the *quark* fields, and where we used the identity:

$$Tr\frac{1}{D}\left|x\right\rangle \Gamma_a \left\langle x\right| = tr\left\langle x\right| \frac{1}{D}\Gamma_a \left|x\right\rangle \tag{8.30}$$

The field correlation function is given by (8.19) with $\Pi = 0$. The graphical expansion of the correlation function is therefore:

$$K = \frac{\delta^2 I(\varphi)}{\delta\varphi\delta\varphi} = -V - VL_0K \tag{8.31}$$

(8.32)

8.6 Next to leading order contribution in N_c

Gap equations and correlation functions which are next to leading order in N_c have recently been evaluated in a self-consistent manner, from the effective action (7.34) in the one-loop approximation [86, 87, 105]. This is equivalent to the symmetry conserving approximation in which the subset of 1-interaction-line irreducible graphs (8.16) consists of the ring graphs:

$\Phi(\varphi) =$ 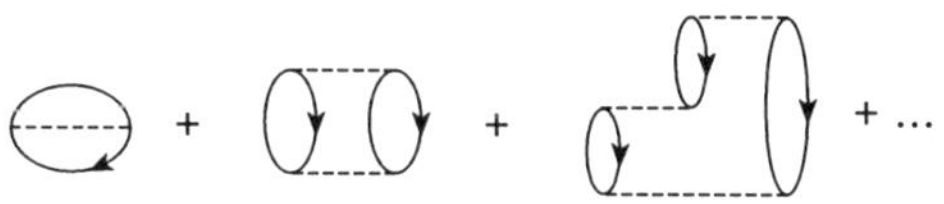

(8.33)

This leads to a symmetry conserving approximation which includes all the graphs which are next to leading order in N_c.

The corresponding generalized gap equation is obtained by setting to zero the first order derivative of the effective action (8.17), with respect to the field $\varphi(x)$:

$$\varphi(x) = -\int d_4y \left\langle x\left|V\right|y\right\rangle \frac{\delta}{\delta\varphi(y)}\left(-Tr\ln D + \Phi(\varphi)\right)$$

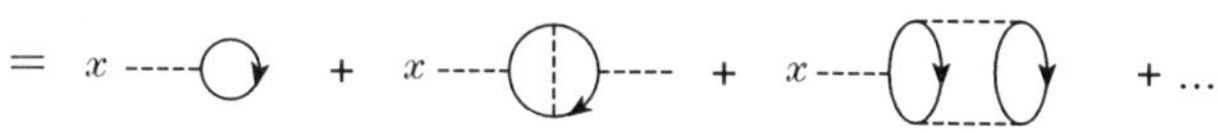

(8.34)

where the wavy line represents the correlation function (8.32) and where we use the abbreviated notation (8.9). At the stationary point of the effective action, the fields are equal to the ground state expectation values of the field operators. Therefore the expression (8.34) is also a graphical representation of the ground state expectation value $\langle \varphi(x) \rangle$.

The inverse correlation function (7.28) can be obtained from the second order derivative of the effective action (8.17):

$$\langle x \left| K^{-1} \right| y \rangle = \frac{\delta^2 \Gamma(\varphi)}{\delta\varphi(x)\,\delta\varphi(y)} = -\langle x \left| V^{-1} \right| y \rangle$$

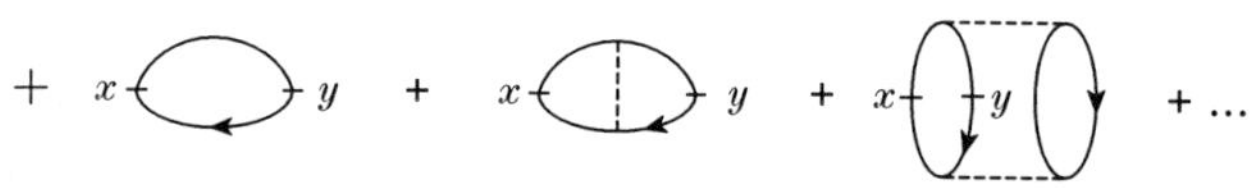

$$= -\langle x \left| V^{-1} \right| y \rangle +$$

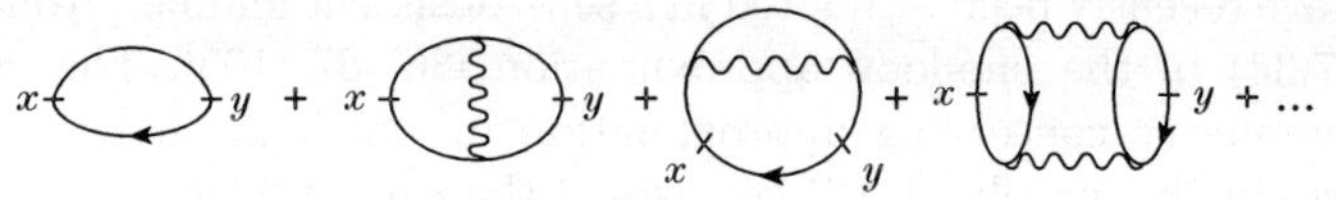

(8.35)

where the wavy line represents the correlation function (8.12). Further applications of next to leading order in N_c contributions can be found in [106, 107, 108].

A problem arises in the evaluation of the right hand sides of equations (8.33), (8.34) and (8.35). They involve the term $\frac{1}{2} Tr \ln \frac{\delta^2 I}{\delta\varphi\delta\varphi}$ and its derivatives, which are ill-defined unless the fields φ are evaluated at a stationary point φ_{class} of the classical action $I(\varphi)$. For other values of the fields, such as the stationary point (8.34) of the effective action $\Phi(\varphi)$, the operator $\frac{\delta^2 I}{\delta\varphi\delta\varphi}$ is not definite positive. A way out of this difficulty is suggested in [87]. It consists in evaluating the right hand sides of equations (8.33), (8.34) and (8.35) at the stationary point φ_{class} of the classical action $I(\varphi)$, *after* the first and second order derivatives have been taken. This procedure maintains the symmetry conserving property and it is consistent with the $1/N_c$ expansion. Indeed, the difference between the evaluation of the right hand side of (8.34) at φ_{class} or at, say, the value of the field φ appearing on the left hand side is reduced by a factor $1/N_c$.

8.7 Self-consistent Schwinger–Dyson and Bethe–Salpeter equations

Symmetry conserving approximations which are distinct from the ones discussed in section 8.3 can be formulated in terms of 2-propagator irreducible graphs. Several works have used these approximations, in connection with effective gluon exchange [20, 74, 109]. For the graph resummations discussed in this section, the reader is referred to [102] and to section 16.2 of [42].

The approximation is defined in terms of 2-propagator irreducible graphs, which cannot be split into two disconnected parts by cutting two quark propagators (oriented lines). Let $\Phi(G)$ be the sum of such 2-propagator irreducible graphs:

$\Phi(G) =$ 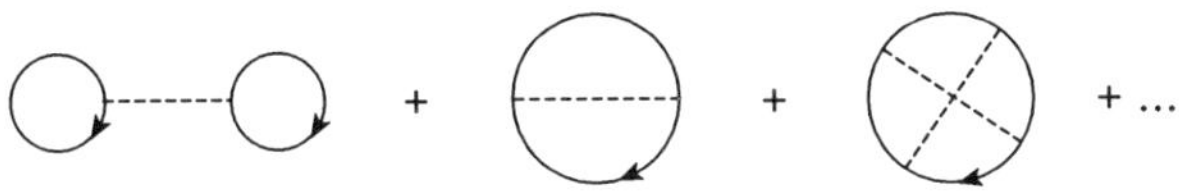 (8.36)

Each graph is to be computed with the rules stated in section 8.2, with the exception that each oriented line, going from a vertex labeled x to a vertex labeled y, contributes a factor $\langle x|G|y\rangle$, where G is a *dressed propagator* defined in equation (8.38), in terms of a self-energy Σ. Any subset of the 2-propagator irreducible graphs (8.36) can be chosen. Each subset defines a symmetry conserving approximation.

To compute the dressed propagator G, we first compute the self-energy Σ, which is defined as the following functional derivative of the irreducible graphs (8.36):

$$\langle 1|\Sigma(G)|2\rangle = \frac{\delta\Phi(G)}{\delta\langle 2|G|1\rangle} =$$ 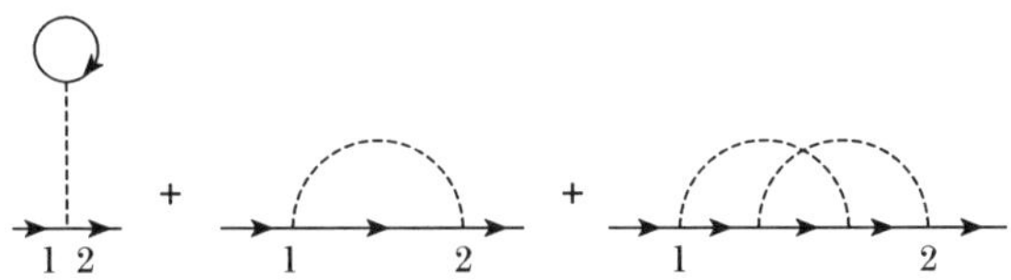 (8.37)

To make the expressions more legible, we denote vertex labels $x \equiv (x, a), y \equiv (y, b), \ldots$ by numbers $(1), (2), \ldots$. It is simple to check that the graphs (8.37) are 1-propagator irreducible, which means that they cannot be split into two disconnected parts by cutting one internal quark propagator. The graphs contributing to $\langle 1|\Sigma(G)|2\rangle$ have one external oriented line entering the labeled vertex (1) and one external oriented line leaving the vertex (2). The internal oriented lines of the graphs (8.37) are also the dressed propagators G, which are computed from the *Schwinger–Dyson equation*:

$$G^{-1} = g^{-1} + \Sigma \qquad G = g - g\Sigma G \tag{8.38}$$

where g is defined in (8.5). The equation (8.38) expresses the dressed quark propagator G as a geometrical series of bare propagators g connected in series by self-energy operators Σ:

$$\langle 1|G|2\rangle = \langle 1|g - g\Sigma g + g\Sigma g\Sigma g - \ldots|2\rangle$$

$=$ 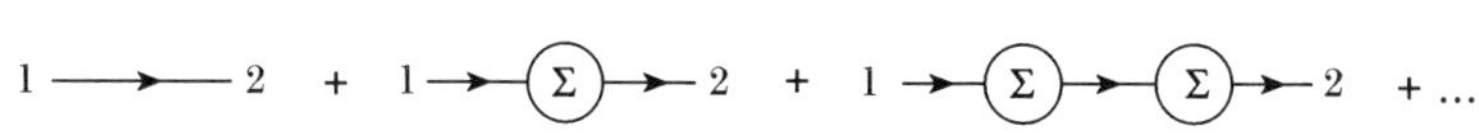 (8.39)

The generating functional (8.10) can be expressed as the following functional of G:

$$\Gamma(G) = -Tr\Sigma G + Tr\ln(1 + g\Sigma) + \Phi(G) \tag{8.40}$$

In spite of the fact that the irreducible graphs $\Phi(G)$ are expressed in terms of the dressed propagators, a topological analysis of the graphs shows that the expression (8.40) includes the contribution of each unlabeled graph (calculated with undressed propagators g) exactly once [42, 102]. The Schwinger–Dyson equation (8.38) can be obtained by writing that the functional $\Gamma(G)$ is stationary with respect to variations of G. We can define a 4-point interaction vertex Π, in the $q\bar{q}$ channel $(1\bar{3})$, by the functional derivative:

$$\left(1\bar{3}\left|\Pi\right|4\bar{2}\right) = \frac{\delta\left\langle 1\left|\Sigma\right|3\right\rangle}{\delta\left\langle 4\left|G\right|2\right\rangle} = \frac{\delta^2\Phi(G)}{\delta\left\langle 3\left|G\right|1\right\rangle\delta\left\langle 4\left|G\right|2\right\rangle} \tag{8.41}$$

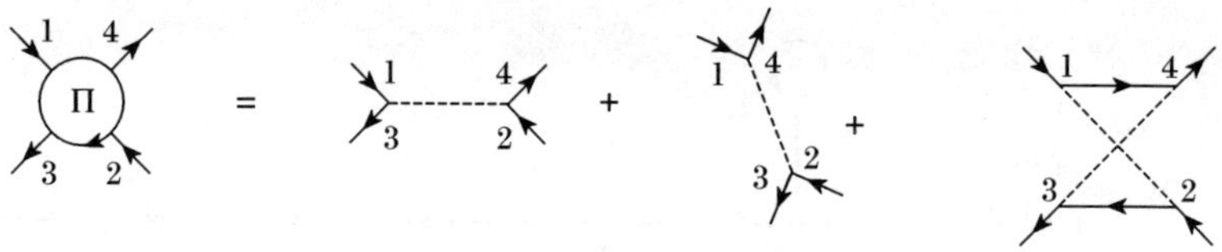

(8.42)

The graphs which contribute to the interaction vertex Π are said to be simple in the $(1\bar{3})$ channel, which means that they cannot be separated into two parts by cutting two propagator lines in that channel. The 4-point function $(13\left|L\right|24)$ can therefore be calculated from the integral equation:

$$\left(1\bar{2}\left|L\right|3\bar{4}\right) = \left\langle 1\left|G\right|3\right\rangle\left\langle 4\left|G\right|2\right\rangle + \left\langle 1\left|G\right|5\right\rangle\left\langle 6\left|G\right|2\right\rangle\left(5\bar{6}\left|\Pi\right|7\bar{8}\right)\left(7\bar{8}\left|L\right|3\bar{4}\right)$$

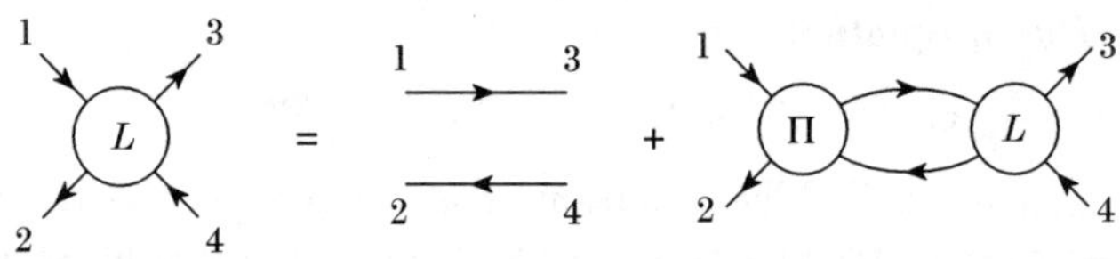

(8.43)

which is called the *Bethe–Salpeter equation.* The graphs (8.43) contribute to the correlation function $(13\left|L\right|24) = \left\langle T\left[(\bar{q}(2)\,q(4))\,(\bar{q}(3)\,q(1))\right]\right\rangle$ which has poles in the (13) channel, corresponding to on-shell meson masses. Self-consistency is obtained when both the operators Σ and Π are calculated as functional derivatives (8.37) and (8.42) of the same subset of 2-propagator irreducible graphs (8.36). When a symmetry is spontaneously broken by the mass operator Σ, the correlation function L acquires a zero eigenvalue, corresponding to a Goldstone boson [42]. This is characteristic of symmetry conserving approximations. The strategy followed here to define a symmetry conserving approximation is quite similar to the one based on the effective action, explained in section 8.3. First, a functional (8.40) is defined in terms of a set of irreducible graphs (8.36), with no external lines. In the effective action formalism, we defined the functional (8.17) of the fields, in terms of the 1-interaction-line irreducible graphs (8.16). In the self-consistent Schwinger–Dyson and Bethe–Salpeter scheme, we defined the functional (8.40) of the quark propagator in terms of the graphs (8.36), which are irreducible with respect to cutting two quark propagators. In both cases, the quark propagator is determined by a stationary point of the action. In both cases the correlation functions are obtained from a second order derivative of the functional, at the stationary point.The two symmetry conserving approximations are however very different, because they correspond to different ways of dressing the quark propagator. To see this, we consider the case where we retain only the following 2-propagator irreducible graph:

$$\Phi(G) = -\int d1d2 \langle 1|G|2\rangle \langle 2|G|1\rangle \langle 1|V|2\rangle = \tag{8.44}$$

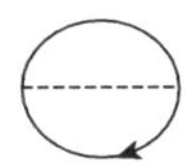

Assume, as in [20, 22, 74, 109], that V represents an effective gluon exchange interaction. Then the direct graph

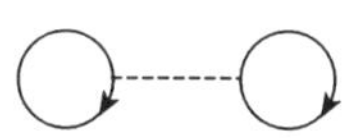

and its first derivative vanish because of the color trace. The second order derivative (8.42) also vanishes in the color singlet channel. The self-energy of the quark propagator is given by (8.37):

$$\langle 1|\Sigma(G)|2\rangle = \frac{\delta\Phi(G)}{\delta\langle 2|G|1\rangle} = -\langle 1|G|2\rangle \langle 1|V|2\rangle = \tag{8.45}$$

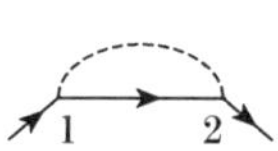

The internal oriented line is a dressed quark propagator G, which is the solution of the Schwinger–Dyson equation (8.38), calculated with the self-energy (8.45). Iteration of the Schwinger–Dyson equation (8.38), with the quark self-energy (8.45), reveals that this internal quark propagator sums the so-called *rainbow graphs*:

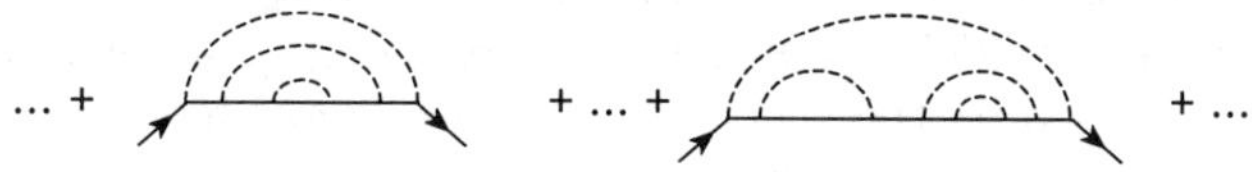

(8.46)

(In the explicit graphs (8.46), the internal oriented lines are the unperturbed propagators g.) The interaction vertex (8.42) reduces to the two-point function:

$$(1\bar{3}\,|\Pi|\,4\bar{2}) = \langle 1 \mid 4\rangle \langle 2 \mid 3\rangle \langle 1\,|V|\,3\rangle =$$

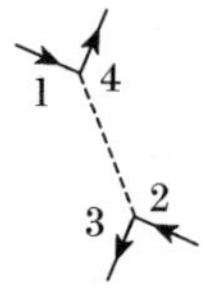

(8.47)

In a color-singlet channel, the Bethe–Salpeter equation (8.43), calculated with the interaction vertex (8.47), sums the ladder graphs:

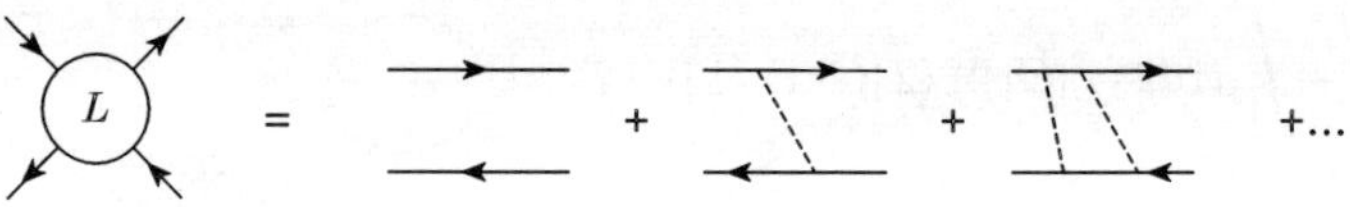

(8.48)

It may be shown that when a symmetry is broken by the quark self-energy Σ, the $q\bar{q}$ correlation function L has a pole corresponding to a zero mass Goldstone boson [42]. The use of 2-propagator irreducible graphs has been applied to the case where the interaction is an effective gluon exchange interaction [20, 74, 109]. In that case, the first irreducible graph in (8.36) vanishes because of the color trace. So do the first graphs in (8.45) and (8.42), at least in the colorless $q\bar{q}$ channel. In [109], the self-consistent scheme includes the second and third 2-line irreducible graphs, displayed in (8.36).The symmetry conserving approximations developed in sections 8.3 and 8.7 correspond to completely different graph re-summations. The graph (8.44) gives rise to completely different correlation functions according to whether it is considered as a-1-interaction-line irreducible graph (8.16) or a 2-propagator irreducible graph (8.36). The symmetry conserving approximation based on the 2-propagator irreducible graphs (8.36) is more difficult to implement numerically. It is richer because it sums ladder graphs, which can induce the proper correlations in the case of strong or singular interactions V.

9

CORRELATION FUNCTIONS IN THE NAMBU JONA-LASINIO MODEL

The reader is advised to go through the calculations in chapter 5 if he wishes to familiarize himself with the techniques applied in this chapter. Reviews of the calculation of correlation functions in the Nambu Jona-Lasinio model can be found in [29, 52]. In [98], euclidean correlation functions are compared with 'experimental' ones deduced from data by dispersion relations. We follow the methods developed in [30].

9.1 Second order expansion of the action

Let us start with the Nambu Jona-Lasinio action (6.5) regularized by delocalized fields. We omit vector fields, which are discussed in section 9.3.1. To calculate the field propagators (7.28) in the vacuum, we expand the effective action in powers of the fields, in the vicinity of a translationally invariant stationary point, for which we guess the diagonal form (5.7). We neglect the second term of (7.34) which is next to leading order in N_c, in which case the effective action reduces to the action (6.5). We redefine the action by subtracting its value at the stationary point and we write it in the form:

$$I(S,P) = -Tr\ln(1+gv) + \frac{1}{4g_s^2} tr_f \int d_4x \left(S^2+P^2-M^2\right) \tag{9.1}$$

where:

$$\begin{aligned} g &= \frac{r^2}{-i\partial_\mu\gamma_\mu + m + r^2M} = G\left(i\partial_\mu\gamma_\mu + m + r^2M\right) \\ G &= \frac{r^2}{-\partial_\mu^2 + \left(m+r^2M\right)^2} \end{aligned} \tag{9.2}$$

is the unperturbed quark propagator and where the interaction:

$$v = (S-M) + i\gamma_5 P \tag{9.3}$$

is first order in the fluctuating parts of the dynamical fields. The regulator r is assumed diagonal in momentum space, so that $\langle k|r|k'\rangle = \delta_{kk'} r\left(k^2\right)$, and $r\left(k^2\right)$ decreases at least as fast as k^2.

To obtain the first order variation of the action, we calculate $Trgv$ in the plane wave basis (3.31):

$$Trgv = TrG\left(i\partial_\mu\gamma_\mu + m + r^2 M\right)\left((S-M) + i\gamma_5 P\right)$$

$$= 4N_c \sum_{i=u,d,s,\dots} \left(\frac{1}{\beta\Omega}\sum_k \frac{r^2\left(m_i + r^2 M_i\right)}{k^2 + \left(m_i + r^2 M_i\right)^2}\right)\int d_4x \left\langle i\left|S-M\right|i\right\rangle \tag{9.4}$$

Using (9.4), it is then simple to calculate the first order variation of the action (9.1) which we set equal to zero. We obtain the *gap equations*:

$$\frac{1}{8N_c g_s^2} M_i = \frac{1}{\beta\Omega}\sum_k \frac{r^2\left(k^2\right)\left(m_i + r^2\left(k^2\right)M_i\right)}{k^2 + \left(m_i + r^2\left(k^2\right)M_i\right)^2} \tag{9.5}$$

These equations differ from the gap equations (5.14) only by the presence of the regulator $r\left(k^2\right)$ which ensures the convergence of the sum over k, and by the fact that we did not make the substitution $m + S \to S'$.

In the chiral limit $m \to 0$, the trivial solution $M = 0$ exists. We are however more interested in the case of spontaneously broken chiral symmetry, which occurs when the coupling constant g_s^2 is strong enough to make the solution $M > m$ energetically favored. We use (9.4) and the gap equations (9.5) so as to express the action (9.1) in a form which is explicitly second order in the fluctuating parts $\widetilde{S} = S - M$ and P:

$$I(S,P) = -Tr\ln(1+gv) + Trgv + \frac{1}{4g_s^2} tr_f \int d_4x \left((S-M)^2 + P^2\right) \tag{9.6}$$

The form (9.6) justifies a posteriori the form (5.7) assumed for the stationary point.

We expand the action (9.6) to second order in $\widetilde{S} = S - M$ and P. The traces eliminate the cross-terms. A straightforward calculation leads to the following second order expansion:

$$I(S,P) = I^{SS} + I^{PP} \tag{9.7}$$

with:

$$\begin{aligned} I^{SS} &= \frac{Tr}{4} Gr^2\left(\partial_\mu \widetilde{S}\right) Gr^2\left(\partial_\mu \widetilde{S}\right) \\ &+ \frac{Tr}{2} G\left(m + r^2 M\right)\left[\left(m + r^2 M\right)\widetilde{S} + \widetilde{S}\left(m + r^2 M\right)\right] G\widetilde{S} \\ &- \frac{Tr}{2} r^2 \widetilde{S} G \widetilde{S} + \frac{1}{4g_s^2} tr_f \int d_4x \widetilde{S}^2 \end{aligned} \tag{9.8}$$

$$\begin{aligned} I^{PP} &= \frac{Tr}{4} Gr^2 (\partial_\mu P) Gr^2 (\partial_\mu P) \\ &+ \frac{Tr}{2} G \left(m + r^2 M\right) \left[\left(m + r^2 M\right) P - P \left(m + r^2 M\right)\right] GP \\ &- \frac{Tr}{2} r^2 PGP + \frac{1}{4g_s^2} tr_f \int d_4 x P^2 \end{aligned} \tag{9.9}$$

The only difference between I^{SS} and I^{PP} is the relative sign in the second term.

9.2 Meson propagators in the scalar channels

The somewhat complicated expressions (9.8) and (9.9) have been derived from the regularized action (9.1). Unfortunately, meson propagators have not been systematically calculated using non-local regularization. One reason is that the determination of on-shell meson masses requires an ill-defined analytic continuation of meson propagators. Some authors proceed in a different way. They derive formal expressions for the propagators, starting from an unregularized action. The divergences which turn up are subsequently regularized. This procedure leads to simpler expressions. The danger is however that correlation functions obtained as functional derivatives may then not be consistent with the existence of an action. Other quantities, such as quark condensates, meson loop contributions, etc., can reveal new divergences which must be regularized in turn.

In order to illustrate one of the most commonly used regularizations, we calculate the propagators in the proper-time regularization discussed in section 6.3.1. The proper-time regularization has the advantage of allowing an analytic continuation of the calculated propagators to time-like momenta, as required for the calculation of on-shell meson masses (section 7.2.1). The analytic properties of proper-time regulated propagators are studied in [85]. We start from the proper-time regularized action (6.18). We make the substitution $S + m \to S$ and we neglect the vector fields:

$$I(S, P) = \frac{1}{2} Tr \int_{\frac{1}{\Lambda^2}}^{\infty} \frac{ds}{s} e^{-sD^\dagger D} + \frac{1}{4g_s^2} tr_f \int d_4 x \left((S - m)^2 + P^2 \right) \tag{9.10}$$

The fields S and P are matrices in flavor space of the form (5.48). The operator $D^\dagger D$ is given by (5.6). To calculate the field propagators, we expand the action (9.10) to second order about the point (5.7). To this order, the imaginary part of the action does not contribute. We write:

$$D^\dagger D = D_0^\dagger D_0 + V \qquad V = S^2 + P^2 - M^2 + i\gamma_5 [S, P] + i\gamma_\mu (\partial_\mu S) + \gamma_5 \gamma_\mu (\partial_\mu P) \tag{9.11}$$

The operator $D_0^\dagger D_0$ is diagonal in flavor and momentum space:

$$D_0^\dagger D_0 |ki\rangle = \left(k^2 + M_i^2\right) |ki\rangle \tag{9.12}$$

where $i = u, s, d, ...$ is a flavor index. The Dyson expansion:

$$e^{-s(A+B)} = e^{-sA}\left(1 - \int_0^s duB(u) + \int_0^s du \int_0^u dvB(u)B(v) + ...\right),$$
$$B(u) \equiv e^{sA}Be^{-sA} \tag{9.13}$$

can be used to expand the action (9.10) in powers of V, with the result:

$$I(S,P) = \frac{1}{2}Tr\int_{\frac{1}{\Lambda^2}}^{\infty}\frac{ds}{s}e^{-sD_0^\dagger D_0}\left(1 - sV + \frac{s^2}{2}\int_0^1 dze^{szD_0^\dagger D_0}Ve^{-szD_0^\dagger D_0}V + ...\right)$$

$$+\frac{1}{4g_s^2}tr_f\int d_4x\left((S-m)^2 + P^2\right) \tag{9.14}$$

The interaction V contains both first and second order terms in the fluctuating parts of the fields $\widetilde{S} = S - M$ and P. We obtain the proper-time regularized gap equation by setting to zero the first order variation of the action:

$$\frac{1}{8N_c g_s^2}\left(1 - \frac{m_i}{M_i}\right) = \frac{1}{\beta\Omega}\sum_k \int_{\frac{1}{\Lambda^2}}^{\infty} ds\ e^{-s\left(k^2+M_i^2\right)} \tag{9.15}$$

The gap equations (9.15) and (5.14) would coincide in the limit $\Lambda \to \infty$.

We can use the gap equation to derive an expression for the action which is the analogue of (5.17):

$$I(S,P) = \frac{Tr}{4}\int_{\frac{1}{\Lambda^2}}^{\infty} sds\left(\int_0^1 dze^{-s(1-z)D_0^\dagger D_0}Ve^{-szD_0^\dagger D_0}V + ...\right)$$
$$+ \frac{1}{4g_s^2}tr_f\frac{m}{M}\int d_4x\left((S-M)^2 + P^2\right) \tag{9.16}$$

It is quadratic in the fluctuating parts of the fields and, in the limit $\Lambda \to \infty$, it has only a logarithmic divergence. It is now straightforward to expand the action to second order in $\widetilde{S} = S - M$ and P. The result has the form (9.7) and we find:

$$I^{SS} = frac12\beta\Omega\sum_q \frac{1}{2}\sum_{ij}\widetilde{S}_{ij}(q)\widetilde{S}_{ji}(-q)\left[4N_c f_{ij}(q)\left(q^2 + (M_i + M_j)^2\right)\right.$$
$$\left. + \frac{1}{2g_s^2}\left(\frac{m_i}{M_i} + \frac{m_j}{M_j}\right)\right]$$
$$I^{PP} = frac12\beta\Omega\sum_q \frac{1}{2}\sum_{ij}P_{ij}(q)P_{ji}(-q)\left[4N_c f_{ij}(q)\left(q^2 + (M_i - M_j)^2\right)\right.$$
$$\left. + \frac{1}{2g_s^2}\left(\frac{m_i}{M_i} + \frac{m_j}{M_j}\right)\right] \tag{9.17}$$

where $f_{ij}(q)$ is the proper-time regularized loop integral:

$$f_{ij}(q) = \frac{\pi^2}{(2\pi)^4} \int_{\frac{1}{\Lambda^2}}^{\infty} sds \int_0^1 dz e^{-s\left((1-z)M_i^2 + zM_j^2 + q^2 z(1-z)\right)} \tag{9.18}$$

and where:

$$\widetilde{S}(q) = \int d_4x \; e^{iq_\mu x_\mu} \widetilde{S}(x) \qquad \int d_4x \; \widetilde{S}(x) S(x) = \frac{1}{\beta\Omega} \sum_q \widetilde{S}(q) \widetilde{S}(-q) \tag{9.19}$$

The function (9.18) diverges if:

$$-q^2 > (M_i + M_j)^2 \tag{9.20}$$

In section 7.2.1, we saw that an on-shell mass M_{mes} corresponds to a pole of the propagator at $-q^2 = M_{mes}^2$. The inequality (9.20) sets a limit $M_{mes} < (M_i + M_j)$ to the on-shell mass which can be calculated in the model. This is a limitation of the model, due to its inability to confine quarks. Indeed, if a meson has a mass $M_{mes} > (M_i + M_j)$, then it can decay into two free quarks of mass M_i and M_j. The corresponding $q\bar{q}$ excitation is unbound. This gives the meson an unphysical width. The action (9.17) shows that, in principle, scalar mesons are all unbound. However, if their mass is close to the threshold $(M_i + M_j)$, then the unphysical width can be quite small.

9.2.1 *Correlation functions of the pseudoscalar nonet*

If we make the matrix P_{ij} explicit as in (5.48), we obtain the second order action I^{PP} in the form:

$$\begin{aligned} I^{PP} = & \frac{1}{2\beta\Omega} \sum_q \sum_{a=1}^{3} P_a(q) P_a(-q) K_\pi^{-1}(q) + \frac{1}{2\beta\Omega} \sum_q \sum_{a=4}^{7} P_a(q) P_a(-q) K_K^{-1}(q) \\ & + \frac{1}{2\beta\Omega} \sum_q P_0'(q) P_0'(-q) K_\eta^{-1}(q) + \frac{1}{2\beta\Omega} \sum_q P_8'(q) P_8'(-q) K_{\eta'}^{-1}(q) \\ & + \frac{1}{2\beta\Omega} \sum_q 8\kappa^2 \left(\frac{\sqrt{2}}{M} P_0'(q) - \frac{1}{M_s} P_8'(q) \right) \left(\frac{\sqrt{2}}{M} P_0'(-q) - \frac{1}{M_s} P_8'(-q) \right) \end{aligned} \tag{9.21}$$

We have added a $U_A(1)$ symmetry breaking term (5.52). The inverse propagators are equal to:

$$K_\pi^{-1}(q) = 4N_c f(q) q^2 + \frac{m}{g_s^2 M}$$

$$K_K^{-1}(q) = 4N_c f_s(q) \left(q^2 + (M - M_s)^2 \right) + \frac{1}{2g_s^2} \left(\frac{m}{M} + \frac{m_s}{M_s} \right)$$

$$K_{\eta'}^{-1}(q) = 4N_c f(q) q^2 + \frac{m}{g_s^2 M} \qquad K_{\eta}^{-1}(q) = 4N_c f_{ss}(q) q^2 + \frac{m_s}{g_s^2 M_s} \tag{9.22}$$

We define:

$$f(q) = f_{uu}(q) = f_{dd}(q) \qquad f_s(q) = f_{us}(q) = f_{ds}(q) \tag{9.23}$$

The π and K on-shell meson masses m_π and m_K are deduced from the zeros of the inverse propagators:

$$m_\pi^2 = \frac{m}{4N_c g_s^2 f(q^2 = -m_\pi^2) M} \tag{9.24}$$

$$m_K^2 = \frac{1}{4N_c g_s^2 f_s(q^2 = -m_K^2)} \frac{1}{2}\left(\frac{m}{M} + \frac{m_s}{M_s}\right) + (M - M_s)^2 \tag{9.25}$$

If we replace the functions $f(q^2 = -m^2)$ by the functions $f(q^2 = 0)$ in (9.25), we recover the values (5.55) of the equivalent linear sigma model.

The on-shell masses which are given by the equations (9.25) require an analytic continuation of the propagators to negative q^2. This is possible in the proper-time regularization, provided that $-q^2 < 4M^2$ (see the discussion following equation (9.18)). The extra singularities, introduced by the proper-time regulator [85], do not lie on the real axis of the complex q^2 plane. An analytic continuation from $q^2 > 0$ to $q^2 < 0$ can therefore be made along the real axis of the q^2 variable. When 4-momentum non-local regulators are used (see section 6.1.1) the analytic continuation, if at all possible, is far from trivial.

The functions K_η^{-1} and $K_{\eta'}^{-1}$ are not the inverse propagators of the η and η' physical mesons, because the $U_A(1)$ symmetry breaking term mixes the fields P_0' and P_8'. To obtain the η and η' propagators, the quadratic form in $P_0'(q)$ and $P_8'(q)$ needs to be diagonalized for each q. This is readily done numerically. Results obtained by various groups are listed in Table 10.1.

In the chiral limit $m \to 0$, and in the absence of the $U_A(1)$ symmetry breaking term, the propagators would all have poles at $q^2 = 0$. There would be nine Goldstone bosons.

9.2.2 *The decay constants f_π and f_K*

The pion and kaon decay constants are defined by the equations:

$$\langle 0 | j_a^\mu(x) | \pi_a, \boldsymbol{q}\rangle = -iq_\mu \frac{e^{iq_\mu x_\mu}}{\sqrt{2\omega_q \Omega}} f_\pi \qquad \langle 0 | j_a^\mu(x) | K_a, \boldsymbol{q}\rangle = -iq_\mu \frac{e^{iq_\mu x_\mu}}{\sqrt{2\omega_q \Omega}} f_K \tag{9.26}$$

where $j_a^\mu(x)$ is the axial current and where $\omega^2 = \boldsymbol{q}^2 + m^2$. The axial current is the Noether current associated to the chiral transformation (2.3). In a local chiral transformation, the Dirac operator transforms to:

$$D \to e^{-i\gamma_5\theta_a(x)\tau_a} D e^{-i\gamma_5\theta_a(x)\tau_a} = D + \gamma_\mu\gamma_5 (\partial_\mu\theta_a) \frac{\tau_a}{2} + ... \tag{9.27}$$

so that the axial current operator is:

$$j_a^\mu(x) = \frac{\delta I}{\delta(\partial_\mu\theta_a)} = \bar{q}\gamma_\mu\gamma_5 (\partial_\mu\theta_a) \frac{\tau_a}{2} q \tag{9.28}$$

Corrections to this expression arise when the Noether current is calculated in the presence of non-local interactions (see the discussion in section 9.4). To calculate the decay constants (9.26) we proceed as in section 7.5. We add a source term $\eta_{\mu a}\gamma_\mu\gamma_5\frac{\tau_a}{2}$ to the Dirac operator D and we use the equation (7.46). The calculation is straightforward and yields:

$$f_\pi^2 = 4M^2 N_c \frac{d}{dq^2}\left(q^2 f\left(q^2\right)\right)\bigg|_{q^2=-m_\pi^2}$$

$$f_K^2 = (M+M_s)^2 N_c \frac{d}{dq^2}\left(q^2 f_s\left(q^2\right)\right)\bigg|_{q^2=-m_K^2} \tag{9.29}$$

The weak decay constants are further discussed in section 9.3.3.

9.2.3 *The quark condensates*

From the action (9.10) we derive the following expressions for the quark condensates:

$$\langle\bar{u}u\rangle = \langle\bar{d}d\rangle = -\frac{M-m}{2g_s^2} \qquad \langle\bar{s}s\rangle = -\frac{M_s-m_s}{2g_s^2} \tag{9.30}$$

The equations (9.25), (9.29) and (9.30) can be combined in various ways to obtain, for example, Gell-Mann Oakes Renner type formulae such as (5.35). Values are given in Table 10.1.

9.2.4 *Correlation functions of the scalar nonet*

Using the explicit form (5.48) of the flavor matrix S_{ij}, the second order action of the scalar mesons, defined in (9.17), is:

$$I^{SS} = \frac{1}{2\beta\Omega}\sum_q\sum_{a=1}^{3}\widetilde{S}_a(q)\widetilde{S}_a(-q)K_{a_0}^{-1}(q) + \frac{1}{2\beta\Omega}\sum_q\sum_{a=4}^{7}\widetilde{S}_a(q)\widetilde{S}_a(-q)K_{K_0^*}^{-1}(q)$$

$$+\frac{1}{2\beta\Omega}\sum_q\widetilde{S}'_0(q)\widetilde{S}'_0(-q)K_{f_0}^{-1}(q) + \frac{1}{2\beta\Omega}\sum_q\widetilde{S}'_8(q)\widetilde{S}'_8(-q)K_{f'_0}^{-1}(q) \tag{9.31}$$

We identify the inverse propagators:

$$K_{a_0}^{-1}(q) = 4N_c f(q)\left(q^2 + 4M^2\right) + \frac{m}{g_s^2 M}$$

$$K_{K_0^*}^{-1}(q) = 4N_c f_s(q)\left(q^2 + (M + M_s)^2\right) + \frac{1}{2g_s^2}\left(\frac{m}{M} + \frac{m_s}{M_s}\right)$$

$$K_{f_0}^{-1}(q) = 4N_c f(q)\left(q^2 + 4M^2\right) + \frac{m}{g_s^2 M} = K_{a_0}^{-1}(q)$$

$$K_{f_0'}^{-1}(q) = 4N_c f_{ss}(q)\left(q^2 + 4M_s^2\right) + \frac{m_s}{g_s^2 M_s} \tag{9.32}$$

We adopted the meson notation of Table 30.2 of the Particle Data [7]. If we set $q = 0$ in the functions $f(q)$ we would recover the masses which can be read off the equivalent linear sigma model (5.56). The conclusions reached in section 5.6.2 still hold when on-shell masses are calculated. Either we adopt constituent quark masses which exceed 500 MeV or we fail to account for the observed masses of the scalar nonet. As argued in section 5.6.2, it is likely that such high constituent quark masses could be avoided if we knew how to add confining forces to the model. Furthermore, as mentioned at the end of section 9.2, the on-shell scalar mesons are all unbound.

9.3 Vector meson propagators

The methods developed in section 9.2 can be applied to the propagators of vector fields, so as to calculate the masses of the $L = 0, S = 1$ mesons $\rho(770)$, $\omega(782)$, $\phi(1020)$ and $K^*(892)$ which form a flavor nonet. The propagators are obtained by expanding the action to second order in the vector fields $V_{\mu a}$ and $A_{\mu a}$. One can start from a regularized action, such as (6.5) or (6.18). Some authors expand the unregularized action and subsequently regularize the loop integrals, at the risk of losing contact with the model. We illustrate this procedure in a simplified $SU(2)_f$ case, involving one isosinglet field S, one isotriplet pseudoscalar field $P_{a=1,2,3}$ associated to the pion, one isosinglet vector field $V_{\mu,a=0}$ associated to the $\omega(782)$, one isotriplet vector field $V_{\mu,a=1,2,3}$ associated to the $\rho(770)$, and one isotriplet axial field $A_{\mu,a=1,2,3}$ associated to the $a_1(1320)$. We refer the reader to [15, 29] for a complete treatment of the $SU(3)_f$ case, which involves a larger number of terms. The exposition given below follows [30].

9.3.1 *Expansion of the action in powers of the vector fields*

To second order, the expansion of the action (4.20) has the form:

$$I^{(2)}(S, P, V, A) = I^{SS} + I^{VV} + I^{PA} \tag{9.33}$$

The term I^{SS} is the $SU(2)_f$ reduction of (9.17):

$$I^{SS}\left(\widetilde{S}\right)=\frac{1}{2\beta\Omega}\sum_q \widetilde{S}(q)\,\widetilde{S}(-q)\left[4N_c f(q)\left(q^2+4M^2\right)+\frac{m}{g_s^2 M}\right] \tag{9.34}$$

The term I^{VV} is:

$$I^{VV}(V)=\frac{1}{2\beta\Omega}\sum_q V_{\mu a}(q)\,V_{\nu a}(-q)\left[4N_c s(q)\left(\delta_{\mu\nu}q^2-q_\mu q_\nu\right)\right.$$
$$\left.+\,\delta_{\mu\nu}\left(\frac{1}{g_v^2}+4N_c t(q)\right)\right] \tag{9.35}$$

The index $a=0,1,2,3$ distinguishes the field $V_{\mu a=0}$ associated to the $\omega(792)$ and the fields $V_{\mu a=1,2,3}$ associated to the three isospin components of the $\rho(770)$. The ω and the ρ become degenerate when equal coupling constants $g_v^2=g_a^2$ are used for the isoscalar and isovector 4-fermion interactions. This is not required by chiral symmetry. The functions $f(q)$, $s(q)$, and $t(q)$ obtained with a sharp 4-momentum cut-off Λ (see section 6.1.1) are:

$$f(q)=\frac{1}{(2\pi)^4}\int_{k<\Lambda} d_4k\int_{-1}^{1}\frac{du}{2}\frac{1}{\left(k^2+M^2+\frac{q^2}{4}\left(1-u^2\right)\right)^2}$$

$$s(q)=\frac{1}{(2\pi)^4}\int_{k<\Lambda} d_4k\int_{-1}^{1}\frac{du}{2}\frac{1-u^2}{\left(k^2+M^2+\frac{q^2}{4}\left(1-u^2\right)\right)^2}$$

$$t(q)=\frac{1}{(2\pi)^4}\int_{k<\Lambda} d_4k\int_{-1}^{1}\frac{du}{2}\left(\frac{k^2+q^2u^2}{\left(k^2+M^2+\frac{q^2}{4}\left(1-u^2\right)\right)}-\frac{2}{k^2+M^2}\right) \tag{9.36}$$

Such expressions would result from the regularized action (6.5), if a step function regulator (6.3) were used. However, this is only true if $q \ll \Lambda$, otherwise the integral over k would have to be limited to the common volume of two 4-dimensional intersecting spheres, separated by a distance q.

The function $4N_c t(q)$ represents the contribution of the quark loop to the vector meson self-energy. It occurs because the 4-momentum cut-off breaks the gauge invariance of the fermion determinant $-Tr\ln D$. In the Dirac operator D, the vector fields $V_{\mu a}$ and $A_{\mu a}$ are minimally coupled to the quarks and the fermion determinant is gauge-invariant. However, strongly interacting vector mesons, such as the $\rho(770)$ and the $\omega(782)$, are not gauge particles. There is therefore no compelling reason to require a gauge-invariant regularization of the quark loop, in the presence of strongly interacting vector fields. Gauge-invariant couplings are of course required for the coupling of electroweak fields which are gauge fields (see section 9.4).

Proper-time regularization preserves the gauge invariance of the quark loop, and when this regularization is used, the function $t(q)$ vanishes. The quark loop does not then contribute to the mass of the vector meson. This does not however make it a gauge particle, because it acquires a self energy $\frac{1}{4N_c s(q) g_v^2}$ from the 4-fermion interaction. The proper-time regularized loop integrals are:

$$f(q) = \frac{\pi^2}{(2\pi)^4} \int_{-1}^{1} \frac{du}{2} \int_{\frac{1}{\Lambda^2}}^{\infty} \frac{ds}{s} e^{-s\left(M^2+\frac{q^2}{4}(1-u^2)\right)}$$

$$s(q) = \frac{\pi^2}{(2\pi)^4} \int_{-1}^{1} \frac{du}{2} \left(1-u^2\right) \int_{\frac{1}{\Lambda^2}}^{\infty} \frac{ds}{s} e^{-s\left(M^2+\frac{q^2}{4}(1-u^2)\right)}, \qquad t(q) = 0 \tag{9.37}$$

The last term I^{PA} involves a mixing of the axial vector field $A_{\mu a}$ with the derivative $(\partial_\mu P_a)$ of the pion field:

$$I^{PA}(P,A) = \frac{1}{2\beta\Omega} \sum_q A_{\mu a}(q) A_{\nu a}(-q) \left[4N_c s(q) \left(\delta_{\mu\nu} q^2 - q_\mu q_\nu\right) \right.$$

$$\left. +\delta_{\mu\nu} \left(\frac{1}{g_v^2} + 4N_c t(q)\right)\right]$$

$$+\frac{1}{2\beta\Omega} \sum_q \left[4M^2 \left(A_{\mu a}(q) + \frac{iq_\mu}{2M} P_a(q)\right) \left(A_{\mu a}(-q) - \frac{iq_\mu}{2M} P_a(-q)\right) 4N_c f(q)\right.$$

$$\left. +\frac{m}{g_s^2 M} P_a(q) P_a(-q)\right] \tag{9.38}$$

In (9.38) the index $a = 1,2,3$. In the chiral limit $m \to 0$, and in the absence of the mass term $\left(\frac{1}{g_v^2} + 4N_c t(q)\right)$, the pion field P_a would get absorbed by the $A_{\mu a}$ field by a Higgs mechanism. Indeed, we could then define a new field $B_{\mu a} = A_{\mu a} - \frac{1}{2M}(\partial_\mu P_a)$, which could be viewed as a gauge transformation of the $A_{\mu a}$ field, and the action (9.38) would describe a massive vector field $B_{\mu a}$.

In general, the fields $A_{\mu a}$ and P_a can be decoupled by defining a new field:

$$B_{\mu a}(q) = A_{\mu a}(q) + \alpha(q) \frac{iq_\mu}{2M} P_a(q)$$

$$\alpha(q) = \frac{16N_c g_v^2 M^2 f(q)}{1 + 4N_c g_v^2 t(q) + 16 N_c g_v^2 M^2 f(q)} \tag{9.39}$$

The action (9.38) then decouples into the two terms:

$$I^{PA}(P,A) = I^{PP}(P) + I^{BB}(B)$$

$$I^{PP}(P) = \frac{1}{2\beta\Omega}\sum_q P_a(q)\, P_a(-q)\left[4N_c f(q)\frac{1+4N_c g_v^2 t(q)}{1+4N_c g_v^2 t(q)+16N_c g_v^2 M^2 f(q)} q^2 + \frac{m}{g_s^2 M}\right]$$

$$I^{BB}(B) = \frac{1}{2\beta\Omega}\sum_q B_{\mu a}(q)\, B_{\nu a}(-q)\left[4N_c s(q)\left(\delta_{\mu\nu}q^2 - q_\mu q_\nu\right) + \delta_{\mu\nu}\left(\frac{1}{g_v^2} + 4N_c t(q) + 16M^2 f(q)\right)\right] \tag{9.40}$$

which describe the physical π and a_1 mesons.

9.3.2 *The $\rho(770)$ and $\omega(782)$ propagators*

The inverse propagators of the vector fields can be read off (9.35). The propagator can be cast into the form:

$$K_{\mu\nu}(q) = \frac{1}{Z(q)\left(q^2 + m_V^2(q)\right)}\left(\delta_{\mu\nu} + \frac{q_\mu q_\nu}{m_V^2(q)}\right)$$

$$m_V^2(q) = \frac{\frac{1}{g_v^2} + 4N_c t(q)}{4N_c s(q)} \qquad Z(q) = 4N_c s(q) \tag{9.41}$$

The functions $s(q)$ and $t(q)$ are defined in (9.36) and (9.37).

The on-shell mass m_ρ^2, which is common to the $\rho(770)$ and the $\omega(782)$, is the pole of the propagator $K_{\mu\nu}(q)$. It is the solution of the equation $m_\rho^2 = m_V^2\left(q^2 = -m_\rho^2\right)$. The functions $f(q)$, $s(q)$ and $t(q)$ become divergent when $-q^2 > 4M^2$. This implies that the constituent quark mass must exceed half the $\rho(770)$ mass:

$$M > 385 \text{ MeV} \qquad m_\rho < 2M \tag{9.42}$$

Otherwise, the $\rho(770)$ or $\omega(782)$ would decay into free quark–antiquark pairs, as discussed following equation (9.18). In other words, for a given value of the constituent quark mass M, the vector coupling constant g_v^2 must be strong enough to bind the $q\bar{q}$ pair, so as to form the ρ meson.

9.3.3 *The pion propagator and f_π*

The pion propagator can be read off (9.40):

$$K_\pi(q) = \frac{1}{Z(q)\left(q^2 + m_P^2(q)\right)}$$

$$m_P^2(q) = \frac{m}{g_s^2 M Z(q)}$$

$$Z(q) = 4N_c f(q) \frac{1 + 4N_c g_v^2 t(q)}{1 + 4N_c g_v^2 t(q) + 16 N_c g_v^2 M^2 f(q)} \tag{9.43}$$

The pion on-shell mass is obtained from the pole of the propagator $K_\pi(q)$. It is given by the equation $m_\pi^2 = m_P^2\left(q^2 = -m_\pi^2\right)$. The observed mass of the $\pi(139)$ can always be fitted by the appropriate choice of the current quark mass m.

The calculation of the pion decay constant f_π is explained in section 9.2.2. The mixing of the P and A fields modifies the expression for f_π only inasmuch as the function $Z(q)$ is modified, as shown in (9.43). The expression for f_π is:

$$f_\pi = M\sqrt{Z\left(q^2 = -m_\pi^2\right)} \tag{9.44}$$

where small terms, proportional to dZ/dq^2, are neglected. The mixing of the P and A fields therefore modifies the expression for the pion decay constant f_π. We shall see in section 10.1 that this in turn modifies significantly the model parameters, as compared to the case where vector fields are not taken into account.

The quark condensate is given by the expression (9.30). If we combine this expression with the expressions for the on-shell pion mass m_π and for the pion decay constant f_π, we find that the Gell-Mann Oakes Renner formula (5.35) remains valid in spite of the mixing between the P and A fields.

9.3.4 *The unbound $a_1(1235)$ meson and the dangers of gradient expansions*

The inverse propagator of the B field, defined in (9.39), can be read off (9.40). The propagator proper can be cast into the form displayed in (9.41) with:

$$m_A^2(q) = m_V^2(q) + 4M^2\frac{f(q)}{s(q)} \qquad Z(q) = 4N_c s(q) \tag{9.45}$$

If we had made a gradient expansion of the action, the expression for the a_1 mass would be obtained by setting $q = 0$ in (9.45). It is simple to check that $\frac{f(q=0)}{s(q=0)} = \frac{3}{2}$. It follows that a gradient expansion of the action would yield the often quoted formula:

$$m_A^2 = m_\rho^2 + 6M^2 \tag{9.46}$$

This result is however completely misleading. Indeed, the on-shell mass m_{a_1} is the solution of the equation $m_{a_1}^2 = m_A^2\left(q^2 = -m_{a_1}^2\right)$. It *always* has a solution such that $m_{a_1}^2 > 4M^2$. This is illustrated in Fig. 9.1. The model fails to bind the $q\bar{q}$ pair in the a_1 channel, because the Higgs mechanism pushes the on-shell a_1 mass above the mass gap $2M$. The gradient expansion predicts a pole of the propagator at $-q^2 = m_A^2$, when in fact no such pole exists. The gradient expansion obviously fails for $-q^2 > 4M^2$. However, even if the model fails to bind the quarks so as to form an a_1 meson, the axial field $A_{\mu a}$ plays an important role, due to its mixing with the pion field P_a. Gradient expansions can be inaccurate even when the model yields a bound $q\bar{q}$ pair. In the ρ channel, for example, the gradient expansion predicts a mass m_ρ which is about 50% higher than the on-shell mass, when $M = 0.4$ GeV [30]. Gradient expansions, as well as the equivalent sigma model discussed in chapter 5, have to be used with caution. They are very useful for understanding the physics but they represent approximations which break down for all processes (including quark loop contributions to coupling constants, decay rates, etc.) which involve excitation energies close to and above the $q\bar{q}$ threshold $2M$. Figure 9.1 shows a graphical solution of the equation $q^2 + m^2(q) = 0$. The curves show the mass operators $m_A^2(q)$, $m_V^2(q)$, $m_S^2(q)$ and $m_P^2(q)$, plotted against $-q^2$. The calculation is made with $M = 0.4$ GeV while fitting f_π and m_π. The point at which these functions intersect the straight line representing $-q^2$ gives the on-shell meson mass squared. The masses $m^2(q)$ can only be calculated for $-q^2 < 4M^2$. Beyond, the functions $f(q)$, $s(q)$ and $t(q)$ diverge.

9.4 Electromagnetic gauge fields in non-local models

The electromagnetic (photon) field a_μ is distinguished from strongly interacting vector fields V_μ in that it couples to a system so as to ensure gauge invariance of the action. This leads to non-trivial couplings when the model is regularized with a 4-momentum cut-off or by the use of delocalized fields (see section 6.1). We illustrate this by a simple example in which the quarks interact with a non-local scalar field, of the form (6.5). Alternative derivations and a more complete discussion can be found in [38, 71]. The quark contribution to the action is:

$$I = \int d_4x d_4y\, \bar{q}(x) \langle x|D|y\rangle \bar{q}(y) \qquad D = -i\partial_\mu\gamma_\mu + rSr \tag{9.47}$$

where r is a regulator which is diagonal in momentum space, as in (6.2). The coupling of the photon field is obtained by making the action invariant in the gauge transformation:

$$q(x) \to e^{-iQ\alpha(x)} q(x) \qquad \langle x|D|y\rangle \to e^{-iQ\alpha(x)} \langle x|D|y\rangle e^{i\alpha Q(y)} \tag{9.48}$$

where Q is the electric charge operator which acts on the flavor variables of the quark field. The action (9.47) can be given a form which is invariant in the gauge transformation (9.48) by making the substitutions:

$$\partial_\mu \to \partial_\mu + iQa_\mu \qquad \langle x|r|y\rangle \to e^{iQ\int_x^y a_\mu(s)ds_\mu} \langle x|r|y\rangle \tag{9.49}$$

The second transformation ensures the connection between the dynamical fields q and S. It vanishes in the local limit $r \to 1$. The gauged Dirac operator is, explicitly:

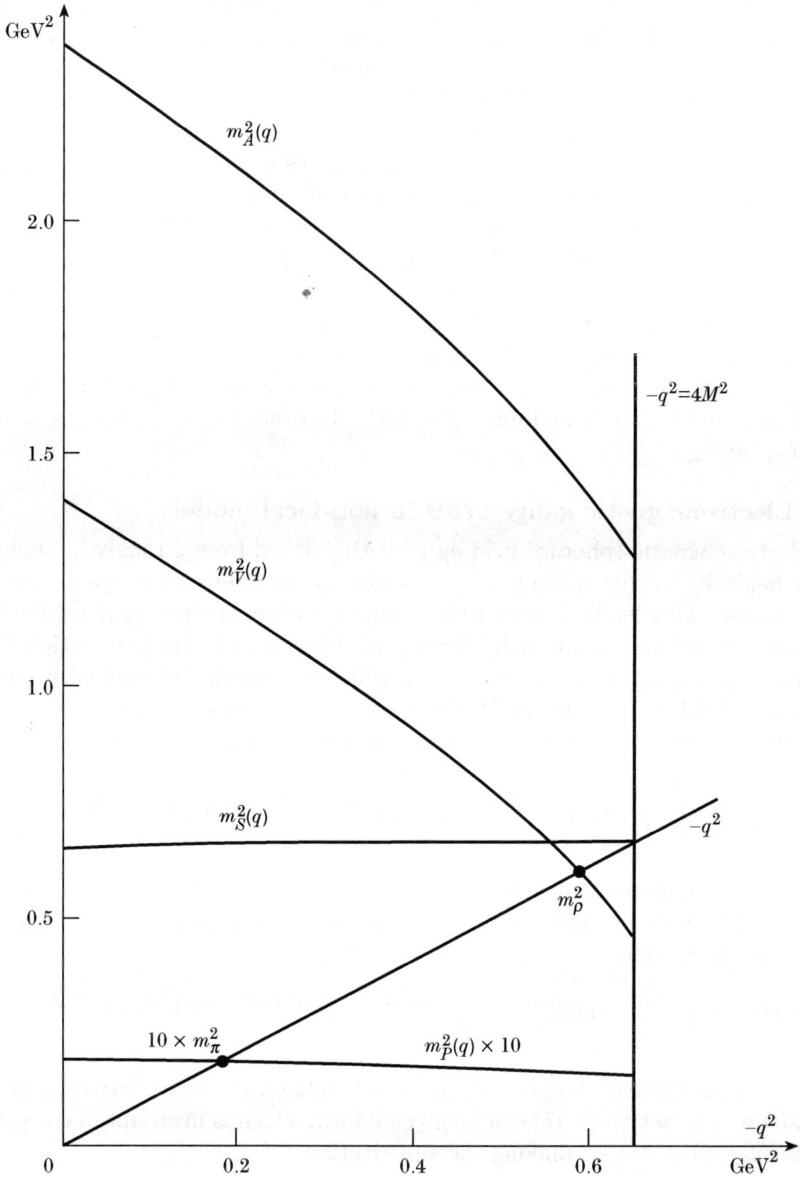

FIG. 9.1. The momentum-dependent meson masses $m^2(q^2)$ are plotted as a function of $-q^2$, where q^2 is the euclidean 4-momentum. An on-shell meson occurs at the point where the curve $m^2(q^2$ intersects the straight line $-q^2$. When $-q^2 > 4M^2$ the loop integrals diverge and the meson is unbound, meaning that it can decay into qq channels.

$$\langle x\,|D|\,y\rangle = -i\partial_\mu\gamma_\mu + Qa_\mu\gamma_\mu + \langle x\,|r|\,z\rangle\, e^{iQ\int_x^z a_\mu ds_\mu} S\,(z)\, e^{iQ\int_z^y a_\mu ds_\mu}\,\langle z\,|r|\,y\rangle \tag{9.50}$$

and the gauged action is $I\,(S,a) = -Tr\ln D$, with D is defined as in (9.50). The gauge invariance of the action $I\,(S,a)$, is expressed by the equations:

$$I\,(S,a) = -Tr\ln D = -Tr\ln e^{-iQ\alpha} De^{iQ\alpha} = I\,(S, a + (\partial\alpha)) \tag{9.51}$$

It follows that $\int d_4x\, \frac{\delta I}{\delta(\partial_\mu\alpha)}\,(\partial_\mu\alpha) = 0$ so that the Noether current $j_\mu = \frac{\delta I}{\delta(\partial_\mu\alpha)}$ is conserved: $\partial_\mu j_\mu = 0$. The contribution of the system to the *photon* self-energy is:

$$\langle x\mu\,|\Pi|\,y\nu\rangle = \frac{\delta^2 I\,(S,a)}{\delta a_\mu\,(x)\,\delta a_\nu\,(y)} = \frac{\delta^2 I\,(S,a)}{\delta\,(\partial_\mu\alpha\,(x))\,\delta\,(\partial_\nu\alpha\,(y))} \tag{9.52}$$

The second equality follows from (9.51). It implies that $\partial_\mu\,\langle x\mu\,|\Pi|\,y\nu\rangle = 0$, from which it follows that the photon remains massless when it interacts with the system. We can display the nature of the couplings which are produced by the non-locality, by evaluating $e^{iQ\int_x^y a_\mu(s)ds_\mu}\,\langle x\,|r|\,y\rangle$. Assume a linear path connecting x and y , which we parametrize as $s_\mu\,(t) = x_\mu + t\,(y_\mu - x_\mu)$. We obtain:

$$\int_x^y a_\mu\,(s)\,ds_\mu = \int_0^1 dta_\mu\,(x + t\,(y-x))\,(y_\mu - x_\mu) \tag{9.53}$$

If the non-locality is weak, that is, if $\langle x\,|r|\,y\rangle$ becomes small for $|x-y| > \frac{1}{\Lambda}$, then we can obtain an expansion in powers of $\frac{1}{\Lambda}$. This involves an expansion of $a_\mu\,(x + t\,(y-x))$ in powers of t. Retaining only the leading order terms in $\frac{1}{\Lambda}$, we find $\int_x^y a_\mu\,(s)\,ds_\mu = a_\mu\,(x)\,(y_\mu - x_\mu)$. It follows that, to leading order in $\frac{1}{\Lambda}$, we have:

$$e^{iQ\int_x^y a_\mu(s)ds_\mu}\,\langle x\,|r|\,y\rangle = e^{-iQa_\mu(x)(x_\mu - y_\mu)+\dots}\,\langle x\,|r|\,y\rangle = \langle x\,|r + Qa_\mu r_\mu \dots|\,y\rangle \tag{9.54}$$

where $r_\mu = \frac{\partial r(k)}{\partial k_\mu}$. The gauged Dirac operator (9.50) becomes, to leading order in $\frac{1}{\Lambda}$:

$$D = -i\partial_\mu\gamma_\mu + Qa_\mu\gamma_\mu + (r + Qa_\mu r_\mu + \dots)\,S\,(r + r_\mu a_\mu Q + \dots) \tag{9.55}$$

We deduce the conserved Noether current:

$$j_\mu\,(x) = \frac{\delta I}{\delta\,(\partial_\mu\alpha)} = -T\left\langle x \left| \frac{1}{D}Q\gamma_\mu + \frac{1}{D}QrSr_\mu + Qr_\mu Sr\frac{1}{D} + \dots \right| x \right\rangle \tag{9.56}$$

In the limit $r \to 1$ of a local theory, we recover the familiar form $j_\mu\,(x) = \langle\bar{q}Q\gamma_\mu q\rangle$ of the Noether current. The expression (9.56) shows that the non-locality contributes extra terms to the conserved current. The latter lead to corrections to the pion decay constant f_π [38] as well as to anomalous processes (see the discussion in section 9.5). Electroweak fields are gauge fields which are used as probes. This means that they are used in expansions of the action in powers of the gauge fields. The coefficients of such expansions can be divergent because the term $Qa_\mu\gamma_\mu$ has no regulator appended to it. Such divergences can be removed by standard gauge-invariant *renormalization.*

9.5 Electroweak fields in proper-time and Pauli–Villars regularized models

The coupling of electroweak fields is simpler in models where only the real part of the action is regularized, using Pauli–Villars or proper-time methods (sections 6.3.2 and 6.3.1). Indeed, these regularizations give rise to local functionals of the fields. As a result, the electroweak fields couple to conserved currents which have the usual form $\bar{q}Q\gamma_\mu q$, where Q is the charge operator. The coupling of electro-weak fields to the system is therefore obtained by making the substitutions $V_{\mu a} \to V_{\mu a} + v_{\mu a}$ and $A_{\mu a} \to A_{\mu a} + a_{\mu a}$ in the Dirac operator (6.18), and *not* in the mass terms. In doing so, we have tacitly made the time components v_{4a} and a_{4a} of the gauge fields purely imaginary. When the fermion determinant $-Tr\ln D$ is expanded in powers of the gauge fields $v_{\mu a}$ and $a_{\mu a}$, the odd powers of the gauge fields contribute to the imaginary part $\frac{1}{2}Tr\ln D^\dagger/D$ of the action, in conjunction with the antisymmetric tensor $\varepsilon_{\mu\nu\alpha\beta}$. This feature depends crucially on using purely imaginary time components of the gauge fields. Terms involving the antisymmetric tensor $\varepsilon_{\mu\nu\alpha\beta}$ give rise to anomalous processes [110, 111, 112], which are known to fit experiment when they are calculated with an infinite cut-off. Indeed, in the case of $\pi \to 2\gamma$, for example, it is known that cut-offs, which can be as low as 800 MeV, can induce up to 40% discrepancies between the calculated and observed values [113]. The fact that anomalous processes require an infinite cut-off to fit experiment provides a phenomenological justification for regularizing only the real part of the action, leaving the imaginary part unregularized. Note however that, when non-local regularizations are used for both the real and the imaginary parts of the action (section 6.1.1), the anomalous processes remain unregularized, provided that one includes the extra terms arising from the non-locality and which are displayed in (9.56). This is proved in [19, 69, 70, 71]. For further discussion of anomalies, see [114].

10

OVERVIEW OF RESULTS IN THE MESON SECTOR

There are several reviews of chiral quark models, such as the Nambu Jona-Lasinio model, applied to the meson sector [29, 52, 115, 116, 117, 118] as well as innumerable applications. Each one stresses a different approach to the subject, although they all boil down to similar calculations. This chapter is an introduction to the vast literature on the subject, and further references appear in the text. We choose a limited number of observables and we expose the physical meaning of the calculations as well as the inevitable problems which are encountered.

10.1 The model parameters and fits to the pseudoscalar nonet

The parameters of the model are:

1. The cut-off Λ. Although different regularizations are used (see chapter 6), they all involve at least one cut-off parameter.
2. The 4-fermion coupling constant g_s^2. It is always chosen strong enough to allow spontaneous breaking of chiral symmetry in the vacuum. This condition is automatically fulfilled if one uses the gap equation to eliminate the coupling constant in favor of one constituent quark mass (usually $M = M_u = M_d$). It is customary to fit the pion decay constant f_π and this provides a relation between the cut-off Λ and the coupling constant g_s^2, or alternatively, a relation between the cut-off and the constituent quark mass M, which remains a free parameter. An example of such a relation is given in Table 5.1.
3. The current quark masses m_i. Unless electromagnetic interactions are also taken into account, it is consistent to take equal current u and d quark masses: $m_u = m_d \equiv m$. They are determined by fitting the pion mass $m_\pi = 138$ MeV. The strange quark mass m_s can be adjusted to fit the kaon mass $m_K = 495$ MeV.
4. The strength κ of the $U_A(1)$ symmetry breaking term. Some authors [29, 51, 52] use the 't Hooft determinant form (4.32) or (4.35). Others [54, 55] use the simpler form (4.37). In either case, the strength κ of the $U_A(1)$ symmetry breaking term is determined by fitting the mass of the $\eta'(958)$. The calculated mass of the $\eta(547)$ can then be compared to experiment.

We list in Table 10.1 some values calculated by five different groups, which we denote by their town of origin:

In the Bochum work, a two-cut-off function was used to regularize the proper-time integral, in order to maintain the non-strange quark condensate $\left\langle \overline{u}u + \overline{d}d \right\rangle^{1/3}$

Table 10.1 *Model parameters (asterisks indicate fitted data)*

Lab	Tübingen	Regensburg	Tsukuba	Bochum	Liège
Reference	[51]	[29]	[52]	[54]	[55]
Regularization	$d_4k < \Lambda$	$d_4k < \Lambda$	$d_3k < \Lambda$	proper-time	$d_4k < \Lambda$
M (MeV)	315	364	335	400	450
M_s (MeV)	508	522	527	606	645
Λ (GeV)	0.805	0.90	0.631		0.721
g_s^2 (GeV)$^{-2}$	12.6	9.80	9.2		24.5
$\langle \overline{u}u + \overline{d}d \rangle^{1/3}$ (MeV) Exp (-283 ± 25)	-274	-311	-308	-299*	-261
$\frac{\langle \overline{s}s \rangle}{\langle \overline{u}u \rangle}$ Exp (0.8 ± 0.1)	0.54	1.13	0.78		1.0
f_π (MeV) Exp (93)	93*	93.3*	93*	93*	93*
f_K (MeV) Exp (114)	100.2	96.3	97.7	105	97.7
m_π (MeV) Exp (139)	140*	139*	138*	139*	139*
m_K (MeV) Exp (114)	496*	498	496*	496*	496*
m_η (MeV) Exp (548)	441	509	495	495	476
$m_{\eta'}$ (MeV) Exp (958)	958*	969	958*	958*	958*

equal to -299 MeV. This makes it difficult to compare the cut-off and the coupling constant g_s^2 with those of the other calculations. The Liège work included a dilaton field (see section 11.2). The numbers quoted are for the case where the dilaton field is essentially frozen. In the $d_4k < \Lambda$ regularization, the 4-momenta of the quark propagators contributing to the loop integrals were cut off at $\left|k_\mu^2\right| < \Lambda$, as done in chapter 5. In the $d_3k < \Lambda$ regularization, the quark 3-momenta were cut off at $\left|\vec{k}\right| < \Lambda$, as in equation (6.14).

Quantities such as the coupling constant g_s^2 and the quark condensates depend quadratically on the cut-off. As mentioned in section 5.5, they are more dependent on the regularization scheme than quantities, such as meson masses, which have only a logarithmic dependence [64]. Table 10.2 shows the different coupling constants and quark condensates obtained from $d_4k < \Lambda$ and proper-time regularization (the pion decay constant is fitted for each constituent quark mass). The results in Table 10.2 are obtained in the chiral limit $m \to 0$ and by fitting $f_\pi = 93$ MeV. The quark condensate can always be fitted by using a two-parameter cut-off function. Note that the quantities which depend quadratically

Table 10.2 *Coupling constants and quark condensates*

M (GeV)	Regularization	Λ (GeV)	$\frac{g_s^2\Lambda^2}{4\pi}$	$\langle\bar{u}u+\bar{d}d\rangle^{1/3}$ (GeV)
0.2	$d_4k < \Lambda$	1.34	0.57	0.37
	proper-time	1.96	0.55	0.48
0.4	$d_4k < \Lambda$	0.74	0.92	0.27
	proper-time	1.29	0.71	0.42
0.8	$d_4k < \Lambda$	0.78	1.77	0.28
	proper-time	1.91	0.86	0.65

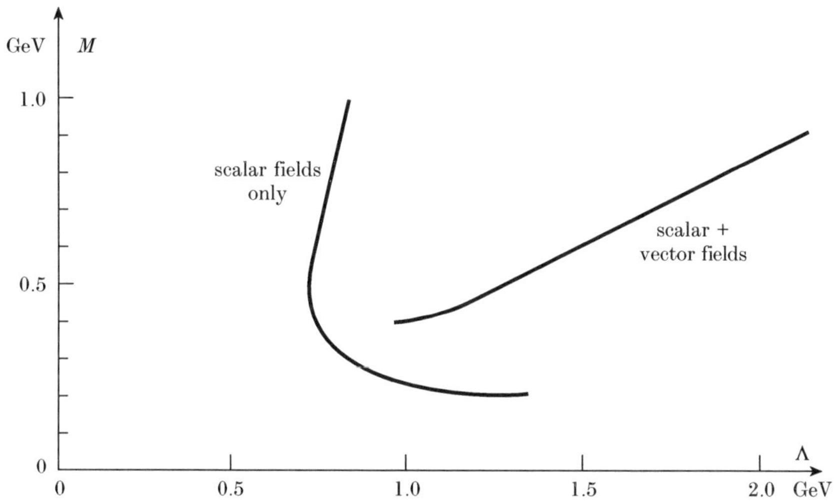

FIG. 10.1. The relation between the constituent quark mass M and the cut-off Λ obtained by fitting f_π. The curve labeled 'scalar + vector fields' is obtained by fitting f_π in the presence of axial vector fields (see section 10.1).

on the cut-off are precisely the ones which have anomalous dimensions in QCD and which are only defined at a given energy scale.

When strongly interacting vector fields are included, the mixing of the pion and axial vector fields modifies the normalization of the pion field, as shown in (9.43). The expression (9.44) of the pion decay constant is therefore modified and so is the relation between the cut-off and the constituent quark mass. This is illustrated in Fig. 10.1, taken from [30].

For a given constituent quark mass M, the mixing of the P and A fields increases the cut-off. As a result, the vacuum becomes more resistant to a restoration of chiral symmetry (see section 10.4).

Table 10.3 *Dependence of vector coupling constant g_v^2 on the type of regularization*

Regularization	Proper-time	$d_4k < \Lambda$
Λ (GeV)	0.90	1.05
$\frac{g_s^2\Lambda^2}{4\pi}$	0.91	0.76
$\frac{g_v^2\Lambda^2}{4\pi}$	0.79	0.55
$\frac{1}{g_v^2}$ (GeV^2)	0.082	0.157
$\frac{1}{g_v^2} + 4N_c t\left(q^2 = -m_\rho^2\right)$ (GeV^2)	0.082	0.105

The vector coupling constant g_v^2 can be determined by fitting the $\rho\,(770)$ mass. The value of the coupling constant depends however on which regularization is used. In section 9.3.2 we saw that the quark loop does not contribute to the ρ and ω masses when the regularization preserves the gauge invariance of the fermion determinant. This is the case of proper-time and Pauli–Villars regularizations. It is not the case of $d_4k < \Lambda$ regularization. This is illustrated in Table 10.3 for a constituent quark mass $M = 0.4$ GeV [30]

The loop integral $t\,(q)$, defined in (9.36), introduces a quadratic dependence of the vector meson mass operator $m_V^2\,(q)$ on the cut-off. However, the quadratic divergence is eliminated from the action when the ρ meson mass is fitted.

10.2 Chiral perturbation theory

Chiral perturbation theory [119, 120] is based on the idea that low energy hadronic processes can be described in terms of the fields which are associated to the lightest particles, namely Goldstone bosons. Spontaneously broken chiral symmetry is implemented by expressing the effective lagrangian in terms of a non-linear representation of the fields such as $U = \exp\left(i\frac{\pi_a\lambda_a}{f_\pi}\right)$, for example. The idea is that, once all the higher energy degrees of freedom are integrated out (from the path integral representing the partition function, for example), one is left with an effective lagrangian, the terms of which are derivatives of U. The lowest order term is the well known Weinberg lagrangian:

$$\mathcal{L} = \frac{f_\pi^2}{4} tr_f \left(\partial_\mu U\right)\left(\partial_\mu U^\dagger\right) + \frac{c}{4} tr_f m \left(U + U^\dagger - 2\right) \tag{10.1}$$

In (10.1), $f_\pi = 93$ MeV is the pion decay constant, c is a phenomenological constant and m is the quark mass matrix. If the Gell-Mann Oakes Renner relation (5.35) is *assumed*, then the constant c is equal to the quark condensate $\left\langle \bar{u}u + \bar{d}d\right\rangle$. Of course, only the product cm can be related to low energy data. Covariant derivatives can be introduced in order to couple electroweak currents to the system. The next order terms contain fourth order derivatives of U and terms which are first order in m and second order in the derivatives of U. At this order, the effective lagrangian has a finite number of terms, the coefficients of which can be determined from a variety of low energy data [121]. Thus the

Table 10.4 *Evaluations of Gasser–Leutwyler coefficients*

	Experiment[120, 121]	Arriola[90]	Bijnens [123]
$10^3 \times L_1$	0.65 ± 0.28	0.96	0.8
$10^3 \times L_2$	1.89 ± 0.26	1.95	1.6
$10^3 \times L_3$	-3.06 ± 0.02	-5.21	-4.1
$10^3 \times L_4$	-0.3 ± 0.5	0	0
$10^3 \times L_5$	2.3 ± 0.2	1.23	1.5
$10^3 \times L_6$	-0.2 ± 0.3	0	0
$10^3 \times L_8$	0.9 ± 0.3	0.62	0.8
$10^3 \times L_9$	7.1 ± 0.3	6.27	6.7
$10^3 \times L_{10}$	-5.6 ± 0.3	-5.42	-5.5

effective lagrangian of chiral perturbation theory can be determined experimentally. Chiral perturbation theory is an effective field theory, the exposition of which is beyond the scope of this book. We recommend the recent lectures of Manohar [122] as a useful introduction to the subject as well as the recent review of Holstein [121].

Chiral perturbation theory is the low energy limit of QCD, as well as of all the other models which implement the symmetries of strong interactions and spontaneous chiral symmetry breaking. It is not a test of QCD, because we do not know how to derive the effective lagrangian from QCD. However, we do know how to derive it from chiral quark models, such as the Nambu Jona-Lasinio model, by integrating out all except the Goldstone fields. The effective lagrangian thus obtained can be compared to the experimentally determined one. This is one way to compare a model to experiment. As an illustration, we quote in Table 10.4 two evaluations of the Gasser–Leutwyler coefficients, which are the coefficients of the effective lagrangian and which are fourth order in the derivatives of U or first order in the current quark masses and second order in the derivatives of U. They are evaluated from the Nambu Jona-Lasinio model, in the presence of vector fields. The reader is referred to [90, 123] for a more complete account

The experimental values are taken from [120]. Updated values from [121] are quoted for L_1, L_2, L_3, L_5, L_9 and L_{10}. The model predicts $L_1 = \frac{1}{2}L_2$ as well as $L_4 = L_6 = 0$. The coefficient L_7 is not quoted, because neither calculation took into account a $U_A(1)$ symmetry breaking term. The Arriola fit requires a constituent quark mass $M = 300$ MeV. This is an independent evaluation of the constituent quark mass.

10.3 Radiative decays of pseudoscalar and vector mesons

A variety of meson decays are mediated by the quark loop term alone. They can test models which describe quarks interacting with chiral fields. The decays are calculated by including electroweak gauge fields in the vector fields of the Dirac operator (4.22). Decays mediated by quark loops include the so-called anomalous processes [124]. For an analysis of chiral anomalies, see [114]. The transition matrix elements listed in Table 10.5 have been calculated by expanding the

Table 10.5 *Experimental and calculated meson decay rates*

Decay	Experiment (KeV) [7] (uncertainty)	Chan[125]	Ebert *et al.*[115]
$\pi^0\,(134) \to 2\gamma$	0.0078 (7%)	0.0073	0.0076
$\eta\,(547) \to 2\gamma$	0.465 (9%)	0.31	0.72
$\eta'\,(958) \to 2\gamma$	4.3 (10%)	5.86	4.2
$\eta'\,(958) \to \rho^0\gamma$	60 (9%)	79	71
$\eta'\,(958) \to \omega\gamma$	6 (13%)	7	7.9
$\rho\,(770) \to e^+e^-$	6.7 (13%)	7.6	
$\rho\,(770) \to \pi^\pm\gamma$	68 (11%)	50	88
$\rho\,(770) \to \eta\gamma$	57 (18%)	31	72
$\omega\,(782) \to e^+e^-$	0.60 (12%)	0.64	
$\omega\,(782) \to \pi^0\gamma$	716 (6%)	741	840
$\omega\,(782) \to \eta\gamma$	7.0 (25%)	8.1	8.0
$\phi\,(1019) \to e^+e^-$	1.37 (3%)	1.42	
$\phi\,(1019) \to \pi^0\gamma$	5.8 (10%)	5.6	5.3
$\phi\,(1019) \to \eta\gamma$	57 (5%)	104	62
$\phi\,(1019) \to \eta'\gamma$	< 2	0.25	0.26
$\phi\,(1019) \to \rho\pi$	571 (6%)	174	
$K^{*0}\,(892) \to K^0\gamma$	116 (9%)	93	140
$K^{*+}\,(892) \to K^+\gamma$	50 (9%)	34	54

quark loop in powers of the gradients of the fields, which is not always a reliable procedure (see section 9.3.4). The table lists decay rates calculated by two groups [115, 125]. The experimental data listed in the table have been updated and taken from the 1994 Particle Data [7].

10.4 Restoration of chiral symmetry at finite baryonic density

When considering systems at finite density and finite temperature, it is useful to exhibit the physical content of the mathematical tools which are used. To study how chiral symmetry is restored at finite density and zero temperature, we can study how the energy of the system varies as a function of the scalar field S. At a stationary point of the effective action, the field S is proportional to the quark condensate, $S = -g_s^2 \langle \bar{q}q \rangle + m$, so that the field $S - m$ can serve as an order parameter for the chiral phase transition. This is certainly true in the framework of the Nambu Jona-Lasinio model, which is discussed below. However, the use of the quark condensate as an order parameter has been questioned [65] and other realizations of chiral symmetry have been proposed [126, 127], which lead to the prediction that vector meson masses should drop as the chiral transition is approached.

10.4.1 *Simple model of spontaneous chiral symmetry breaking*

As a simple illustration of spontaneous chiral symmetry breaking, we consider the $SU\,(2)_f$ case where the action is given by (5.24) and we take the chiral limit

$m \to 0$. In the vacuum, the pion field P vanishes. The quark loop term then simply describes free quarks which have a constituent quark mass S. The energy per unit volume of the system is:

$$\frac{E_0(S)}{\Omega} = -\frac{4N_c}{\Omega} \sum_{|\boldsymbol{k}|<\Lambda} \sqrt{\boldsymbol{k}^2 + S^2} + \frac{1}{2g_s^2} S^2 \tag{10.2}$$

The first term is the energy of a Dirac sea of quarks, limited to states with momentum $|\boldsymbol{k}| < \Lambda$. We use the 3-momentum cut-off regularization discussed in section 6.2. It is equally easy to use other regularizations, except that the expressions become less transparent. The first term decreases with S, linearly at large S. The second term increases quadratically with S. Provided the coupling constant g_s^2 is strong enough, the energy acquires a minimum for a finite value $S = M$, which is the vacuum constituent quark mass. The value of M is given by the gap equation:

$$-\frac{4N_c}{\Omega} \sum_{|\boldsymbol{k}|<\Lambda} \frac{M}{\sqrt{\boldsymbol{k}^2 + M^2}} + \frac{M}{g_s^2} = 0 \tag{10.3}$$

This is the Nambu mechanism for spontaneous chiral symmetry breaking. The gap equation (10.3) is equivalent (in the chiral limit $m \to 0$) to the gap equation (5.14) when a 3-momentum regularization is used.

If we subtract the energy at the minimum, the vacuum energy $E_0(S)$ per unit volume becomes:

$$\frac{E_0(S)}{\Omega} = -\frac{4N_c}{\Omega} \sum_{|\boldsymbol{k}|<\Lambda} \left(\sqrt{\boldsymbol{k}^2 + S^2} - \sqrt{\boldsymbol{k}^2 + M^2} - \frac{S^2 - M^2}{2\sqrt{\boldsymbol{k}^2 + M^2}} \right) \tag{10.4}$$

which has a minimum at $S = M$ as shown in Fig. 10.2, taken from [30].

10.4.2 *Quarks or nucleons in the Fermi sea?*

The system at finite baryonic density can be studied by adding a chemical potential μ to the Dirac hamiltonian, which becomes $h = \frac{\boldsymbol{\alpha}.\boldsymbol{\nabla}}{i} + \beta S + i\gamma_5 P_a \tau_a - \mu$. Since the ground state of a system of free fermions is obtained by filling all the negative energy orbits, the effect of the chemical potential is to fill all the plane wave orbits with energies lower than μ. If $\mu > M$, the ground state of the system becomes a Fermi sea of *quarks,* filling all the positive energy plane wave states with $|\boldsymbol{k}| < k_F$, where:

$$\sqrt{k_F^2 + M^2} = \mu \tag{10.5}$$

Since each quark carries a baryonic number $1/N_c$, the Fermi momentum is related to the baryonic density ρ:

$$\rho = \frac{2}{3\pi^2} k_F^3 \tag{10.6}$$

Normal nuclear matter density is the density which prevails in the center of large nuclei. It is the saturation density of nuclear matter, that is, the density at

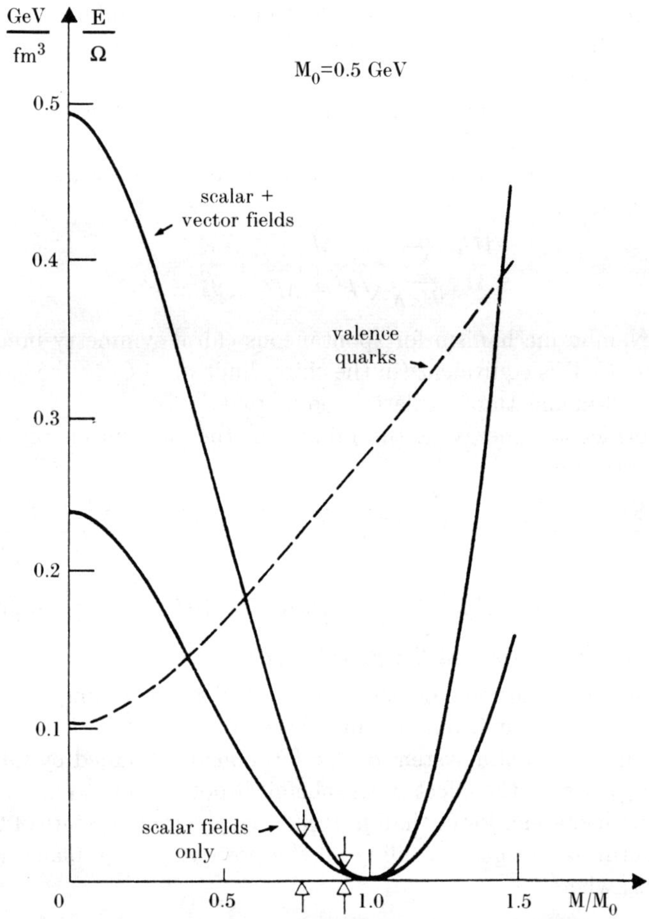

FIG. 10.2. The curve labeled 'scalar fields only' is the energy per unit volume (10.4) of the vacuum as a function of the constituent mass M in units of its vacuum vlaue M_0. The dashed line is the Fermi sea contribution, represented by the first term of (10.7). The curve labeled 'scalar + vector fields' is the vacuum energy obtained by fitting f_π in the presence of axial vector fields (see section 10.1). The arrows show the position of the minimum energy when the contribution of the valence quarks is added.

which symmetric nuclear matter would exist at zero pressure if coulomb forces were switched off. Normal nuclear matter density occurs at $\rho = 0.16$ fm^{-3} and $k_F = 1.3$ fm^{-1} $= 0.26$ GeV. At four times normal nuclear matter density, $\rho = 0.64$ fm^{-3} and $k_F = 2.1$ fm^{-1} $= 0.41$ GeV. The Fermi momentum k_F is thus smaller than the cut-off Λ. The use of a chemical potential is thus tantamount to the description of nuclear matter as a Fermi sea of quarks. In section 12.5 we show that proper-time regularization does not allow the use of a chemical potential. In this, as well as in several other cases, the action is regularized at zero density and an unregularized Fermi sea of quarks is added, as shown below in (10.11).

The energy per unit volume of nuclear matter at finite density can be written in the form:

$$\frac{E\left(S\right)}{\Omega}=\frac{4N_c}{\Omega}\sum_{|\boldsymbol{k}|<k_F}\sqrt{\boldsymbol{k}^2+S^2}+\frac{E_0\left(S\right)}{\Omega} \tag{10.7}$$

The first term, which is the Fermi sea contribution, is a monotonically rising function of S with a slope which increases with baryonic density. It shifts the minimum of $E\left(S\right)$ to a lower value, as illustrated in Fig. 10.2.

As the density increases, the minimum of $E\left(S\right)$ decreases until eventually a minimum occurs at $S = 0$, at which point the quarks lose their constituent mass and chiral symmetry is restored. This is the process by which chiral symmetry is restored at high densities.

So far, we have described nuclear matter as a Fermi sea of quarks. However, below the deconfinement density, nuclear matter is in fact composed of nucleons and not of quarks. This is an inescapable ambiguity of the model. Does it make any difference to the process of chiral symmetry restoration? We can estimate the difference in the Walecka model [128], in which it is assumed that nucleons get their mass from their interaction with the scalar field S. In this model, the nucleon, at finite density, has a mass equal to gS, where g is a coupling constant due to its structure. If $S = M$ is the vacuum value of the scalar field, then the observed value $M_N = 939$ MeV of the nucleon mass is equal to gM. We deduce that, at finite density, the nucleon mass is $\frac{M_N}{M}S$. The nucleons in the Fermi sea contribute an energy per unit volume equal to:

$$\frac{E_F\left(s\right)}{\Omega}=\frac{4}{\Omega}\sum_{|\boldsymbol{k}|<k_F}\sqrt{\boldsymbol{k}^2+\frac{M_N^2}{M^2}S^2} \tag{10.8}$$

(Note that, at a given *baryonic* density ρ, the Fermi momentum k_F is the same for nucleons as for quarks.) Thus, if we assume a Fermi sea of nucleons, rather than of quarks, we must substitute the first term of (10.7) by the expression (10.8). If quarks are weakly bound (as they are in chiral solitons), then the nucleon mass $M_N \approx N_c M$. At normal nuclear matter density, the Fermi momentum is about 0.26 GeV, considerably smaller than the field S. In that case, we can neglect the term k^2 and we find that a Fermi sea of quarks gives roughly the same contribution to the energy as a Fermi sea of nucleons. This is only true for (vacuum) constituent quark masses which are in the 300 – 400 MeV range.

At low densities, it is possible to make a model-independent evaluation of the density dependence of both the quark and gluon condensates [129]. Indeed, since $\langle\bar{q}q\rangle = \frac{\partial E/\Omega}{\partial m}$ we have, at low density where the nucleons are uncorrelated:

$$\langle\bar{q}q\rangle_\rho = \langle\bar{q}q\rangle_0 + \rho\frac{\delta M_N}{\delta m} \tag{10.9}$$

where $\langle\bar{q}q\rangle_\rho$ and $\langle\bar{q}q\rangle_0$ are respectively the quark condensates at finite density ρ and at zero density. The derivative $\frac{\delta M_N}{\delta m}$ is related to the nucleon sigma-term $\sigma_N = m\frac{\delta M_N}{\delta m}$ where m is the current quark mass. The experimental value of σ_N, still controversial, is estimated to be $\sigma_N \approx 45$ MeV [130]. From (10.9) we deduce the model-independent relation:

$$\frac{\langle\bar{q}q\rangle_\rho - \langle\bar{q}q\rangle_0}{\langle\bar{q}q\rangle_0} = \rho\frac{\sigma_N}{m\,\langle\bar{q}q\rangle_0} = -\rho\frac{\sigma_N}{m_\pi^2 f_\pi^2} \tag{10.10}$$

where we used the Gell-Mann Oakes Renner relation $m\,\langle\bar{q}q\rangle_0 = -m_\pi^2 f_\pi^2$, also satisfied by the model, as seen in equation. (5.35). It is quite remarkable that the low density estimate, obtained in the Nambu Jona-Lasinio model for a constituent quark mass $M \approx 320$ MeV, turns out to be close to the value (10.10), in spite of the fact that the model does not explicitly involve the nucleon sigma-term.

The critical density at which chiral symmetry is restored depends strongly on the model parameters and it can vary by a factor of 2 or 3. As a result, the model is unable to predict the critical density with any certainty. It depends strongly on the chosen constituent quark mass M in the vacuum, on whether nucleons or quarks populate the Fermi sea, and on whether f_π is fitted in presence of a_1 vector fields or not [30]. In such a situation, it would seem advisable to fine-tune the model parameters so as to fit the model-independent estimate (10.10), as done in [129], before attempting to predict the critical density. The critical density, derived from (10.10), is $\rho_c \approx m_\pi^2 f_\pi^2/\sigma_N \approx 0.5\ \text{fm}^{-3}$ (about 3 times normal nuclear density). The nucleon sigma term σ_N is unfortunately still quite uncertain. An example of the variation of the quark and gluon condensates, as a function of density, is shown in Fig. 11.1.

One could also argue that, at high density, a phase transition will occur in nuclear matter, when the quark wavefunctions of the nucleons begin to overlap. The relevant parameter becomes then the size of the quark wavefunctions in the nucleon. A description of nuclear matter in terms of a Fermi sea of uncorrelated nucleons or quarks completely neglects such effects.

10.4.3 *Mesons propagating in a dense medium*

Since the constituent quark mass decreases when the baryonic density of the system increases, we expect that hadron masses will also decrease. Indeed, mesons are $q\bar{q}$ excitations and hadrons are bound states of N_c quarks. A meson will polarize the medium in which it is propagating. Chiral quark models can be used to study such processes.

It is convenient to separate the action into two terms:

$$I_{\mu}(S,P,V,A) = I_{\Lambda,\mu=0}(S,P,V,A) + \left[I_{\mu}(S,P,V,A) - I_{\mu=0}(S,P,V,A)\right] \tag{10.11}$$

The first term $I_{\Lambda,\mu=0}(S,P,V,A)$ is the zero density *regularized* action. It could have the form (6.5) or (6.18) for example. The second term is the unregularized Fermi sea contribution. The procedure of not regularizing the Fermi sea contribution should be valid as long as the Fermi momentum k_F remains smaller than the cut-off Λ. This remains true even at densities several times normal nuclear matter density. So long as calculations are limited to one-loop contributions, the second term of (10.11) remains finite. It could however produce ultraviolet divergences if higher order processes were calculated.

The separation of the action (10.11) into two terms is a convenient way to distinguish processes which are caused by the Fermi sea of quarks from the $q\bar{q}$ excitations at zero density ($\mu = 0$), which are responsible for the formation of the mesons in the vacuum. We can use the action (10.11) to calculate the meson propagators (which are the field correlation functions) at finite density, following the method explained in chapter 9. For a given value of μ or k_F, we begin by finding a stationary point of the action (10.11). We then calculate the inverse propagators by expanding the action to second order in the fluctuating parts of the fields. The second term of (10.11) involves all the processes which are linked to the Fermi sea. It leads to more complicated expressions than in the vacuum, because the chemical potential, contained in the constant vector $W_{\mu} = (-i\mu, 0, 0, 0)$, breaks Lorentz covariance. As a result, scalar and vector fields mix with momentum-dependent coefficients and the system displays excitation modes which have different quantum numbers than they do in the vacuum [131, 132].

A Fermi sea of quarks is certainly not a good description of nuclear matter in the confined phase. Processes such as the polarization of the medium and the absorption of mesons in nuclear matter are better described in terms of mesons interacting with nucleons [131, 133, 134]. However, possible modifications of the internal quark structure of a meson propagating in the medium can usefully be studied by chiral quark models.

There is however no unambiguous way to distinguish the two effects. Consider the case where the action is separated into two terms, as in (10.11). The gap equations are solved so as to determine the static values of the fields at finite density, as done in section 10.4. The contribution of the first term $I_{\Lambda,\mu=0}(S,P,V,A)$ to the meson propagators depends on the density because the constituent quark mass decreases with density. This contribution can therefore be interpreted as a change in the intrinsic structure of the meson. Its effect is to reduce all meson

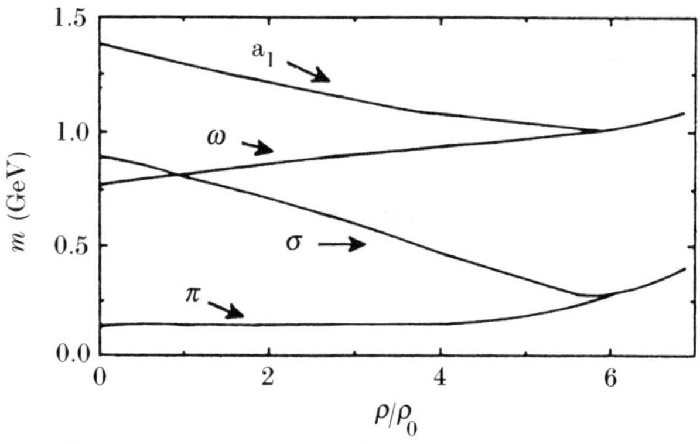

FIG. 10.3. Masses of mesons at rest in the lab frame are plotted as a function of density relative to normal nuclear density ρ_0 From [132].

masses, roughly in proportion to the constituent quark mass, except for the pion, which remains a Goldstone boson even at finite density.

From the second term of (10.11), we can calculate the Fermi sea contribution to the meson propagators. It could be replaced by a term describing the interaction of nucleons (rather than quarks) with the meson fields [135]. The Fermi sea contribution involves excitations of quarks (or nucleons) in the Fermi sea and the excitations of quarks in the Dirac sea. The excitation of Fermi sea quarks (or nucleons) represents the polarization of nuclear matter by the propagating meson, as well as its possible absorption. The excitation of Dirac sea quarks represents a Pauli blocking effect. Indeed the Fermi sea blocks part of the phase space available for $q\bar{q}$ excitations which occur in the vacuum. This may also be viewed as a change in the intrinsic structure of the meson. It is far from clear whether Pauli blocking is better evaluated with a Fermi sea of quarks or nucleons. Pauli blocking is the only effect that arises for a meson at rest in the lab frame. Figure 10.3, taken from [132], shows how the masses of mesons, at rest in the lab frame, vary with density.

Pauli blocking is an important effect. It prevents the ρ and ω masses from decreasing with density [136]. As chiral symmetry becomes restored, the σ and π mesons become degenerate, and so do the ρ (or ω) and a_1 mesons. Quite different behaviour can be obtained when dilaton fields are introduced (section 11.2). Examples of calculations of meson propagators at finite density can be found in Refs.[52, 132, 136, 137, 138, 139].

10.5 Systems at finite temperature

In discussing finite temperature calculations, it is important to distinguish the mathematical tool and the physical content. Models cannot predict the temperature dependence of their input parameters, such as coupling constants and cut-offs. They are kept constant unless some external information is fed in, from lattice calculations for example. So far, we have been evaluating the partition function $Tre^{-\beta H}$ in the limit $\beta \to \infty$ of zero temperature. When considering systems at finite temperature, there is no need to reformulate the field theory. All we need to do is to keep the inverse temperature β finite. This in turn requires a modification of the plane wave states (3.31), so as to implement correctly the boundary conditions of the fields in the path integral.

10.5.1 *The mathematical tool*

When the partition function is expressed in terms of the quark (fermion) field, as in (4.5) or (4.7), the integration 'paths' of the quark field $q(\tau)$ are restricted by the antiperiodic boundary condition [42]:

$$q(\tau = \beta, \boldsymbol{r}) = -q(\tau = 0, \boldsymbol{r}) \tag{10.12}$$

The boundary condition must be respected when the quark field is expressed in momentum space:

$$q(\tau, \boldsymbol{r}) \equiv \langle \tau, \boldsymbol{r} \mid q \rangle = \sum_{k_0, \boldsymbol{k}} \langle \tau \boldsymbol{r} \mid k_0 \boldsymbol{k} \rangle \langle k_0, \boldsymbol{k} \mid q \rangle = \frac{1}{\sqrt{\beta\Omega}} \sum_{k_0, \boldsymbol{k}} e^{i\tau k_0 + i\boldsymbol{k}.\boldsymbol{r}} q(k_0, \boldsymbol{k}) \tag{10.13}$$

The boundary condition (10.12) requires k_0 to take the discrete values:

$$k_0 = \frac{(2n+1)\pi}{\beta} \equiv \omega_n \qquad \text{(for fermion fields)} \tag{10.14}$$

where n is an integer. The ω_n are known as the *Matsubara frequencies* for fermions. When the partition function is expressed in terms of boson fields φ, as in (4.19) or (4.23), the integration 'paths' of the boson fields $\varphi(\tau, \boldsymbol{r})$ are restricted by the periodic boundary condition:

$$\varphi(\tau = \beta, \boldsymbol{r}) = \varphi(\tau = 0, \boldsymbol{r}) \tag{10.15}$$

The Fourier transform $\varphi(q_0, \boldsymbol{q}) = \int_0^\beta d\tau \int_\Omega d_3 r e^{iq_0\tau + i\boldsymbol{k}.\boldsymbol{r}} \varphi(\tau, \boldsymbol{r})$ will satisfy the periodic boundary condition (10.15) if q_0 takes the discrete values:

$$q_0 = \frac{2n\pi}{\beta} \equiv \omega_n \qquad \text{(for boson fields)} \tag{10.16}$$

where the ω_n are the Matsubara frequencies for bosons. When composite operators, such as $S_a(x) = -g_s^2 \bar{q}(x) \tau_a q(x)$ are introduced, as in (4.13) or (4.14), the boundary conditions (10.12) and (10.15) are compatible. However, when dealing with bosonized forms of the action, such as (4.19) and (4.23) for example, it is important to remember that the trace $-Tr \ln D$ is taken over the *quark* variables, so that the trace needs to be expressed in terms of the *fermion* Matsubara frequencies (10.14). This again leads to consistent results, because matrix elements of the boson fields, such as $\langle k |S| k' \rangle = \frac{1}{\beta\Omega} S(k' - k)$, involve only *differences* between fermion Matsubara frequencies, and the latter are boson Matsubara frequencies.The use of Matsubara frequencies is the exact analogue of the use of periodic boundary conditions in a large box of volume $\Omega = L^3$. The only formal change which occurs at finite temperature is the use of the following modified plane-wave states:

$$\langle \tau \mid \omega_n \rangle = \frac{1}{\sqrt{\beta}} e^{i\omega_n \tau} \qquad \langle \boldsymbol{r} \mid \boldsymbol{k} \rangle = \frac{1}{\sqrt{\Omega}} e^{i\boldsymbol{k}.\boldsymbol{r}} \qquad \langle x \mid q \rangle = \frac{1}{\sqrt{\beta\Omega}} e^{ik_\mu x_\mu}$$

$$\boldsymbol{k} = \frac{2\pi}{L}\boldsymbol{n} \qquad k_0 = \omega_n = \begin{cases} \frac{(2n+1)\pi}{\beta} & \text{(fermions)} \\ \frac{2n\pi}{\beta} & \text{(bosons)} \end{cases} \tag{10.17}$$

The plane wave states have orthogonality and completeness relations:

$$\langle \boldsymbol{r}_1 \mid \boldsymbol{r}_2 \rangle = \delta(\boldsymbol{r}_1 - \boldsymbol{r}_2) \qquad \langle \tau_1 \mid \tau_2 \rangle = \delta(\tau_1 - \tau_2) \qquad \langle x_1 \mid x_2 \rangle = \delta(x_1 - x_2)$$

$$\langle \boldsymbol{k}_1 \mid \boldsymbol{k}_2 \rangle = \delta_{\boldsymbol{k}_1,\boldsymbol{k}_2} \qquad \langle \omega_{n_1} \mid \omega_{n_2} \rangle = \delta_{n_1,n_2} \qquad \langle k_1 \mid k_2 \rangle = \delta_{k_1 k_2}$$

$$\int_\Omega d_3 r \,|\boldsymbol{r}\rangle \langle \boldsymbol{r}| = 1 \qquad \int_0^\beta d\tau \,|\tau\rangle \langle \tau| = 1 \qquad \int d_4 x \,|x\rangle \langle x| = 1$$

$$\sum_{\boldsymbol{k}} |\boldsymbol{k}\rangle \langle \boldsymbol{k}| = 1 \qquad \sum_n |\omega_n\rangle \langle \omega_n| = 1 \qquad \sum_k |k\rangle \langle k| = 1 \tag{10.18}$$

In the zero temperature limit $\beta \to \infty$, we can replace the sums by integrals:

$$\sum_{\boldsymbol{k}} \to \frac{\Omega}{(2\pi)^3} \int d_3 k \qquad \sum_{\omega_n} \to \frac{\beta}{2\pi} \int_{-\infty}^{\infty} dk_0 \qquad \sum_k \to \frac{\beta\Omega}{(2\pi)^4} \int d_4 k \tag{10.19}$$

At finite temperature, explicit Lorentz covariance is lost because the time variable is limited to the interval $0 < \tau < \beta$, whereas the space variables are limited to a box of size L which is arbitrarily large. In principle, the use of the plane waves (10.19) is the only change required to calculate the partition function at finite temperature. However, expressions acquire a more familiar form when the sums over Matsubara frequencies are replaced by contour integrals in the complex k_0 plane as follows:

$$\frac{1}{\beta} \sum_n e^{-i\omega_n \tau} f(i\omega_n) = \pm \oint \frac{dz}{2\pi i} \frac{e^{-z\tau}}{e^{\mp\beta z} - 1} f(z) \qquad \omega_n = \frac{2n\pi}{\beta}$$

$$\frac{1}{\beta} \sum_n e^{-i\omega_n \tau} f(i\omega_n) = \pm \oint \frac{dz}{2\pi i} \frac{e^{-z\tau}}{e^{\mp\beta z} + 1} f(z) \qquad \omega_n = \frac{(2n+1)\pi}{\beta} \tag{10.20}$$

where the $+$ sign is chosen when $\tau > 0$ and the $-$ sign when $\tau < 0$. The integration contour, shown in Fig. 10.4(a), can usually be replaced by the contour

(a)

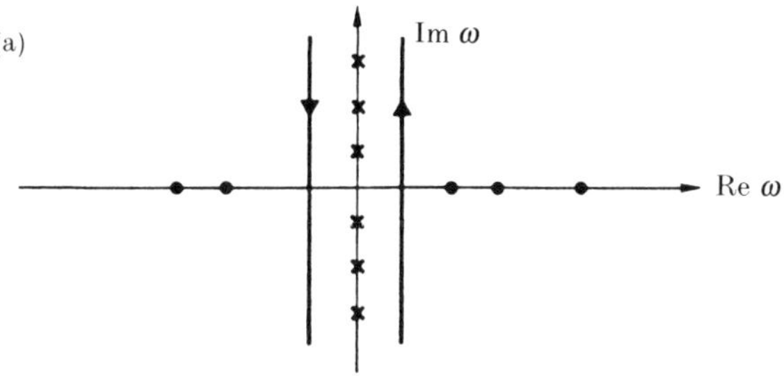

(b)

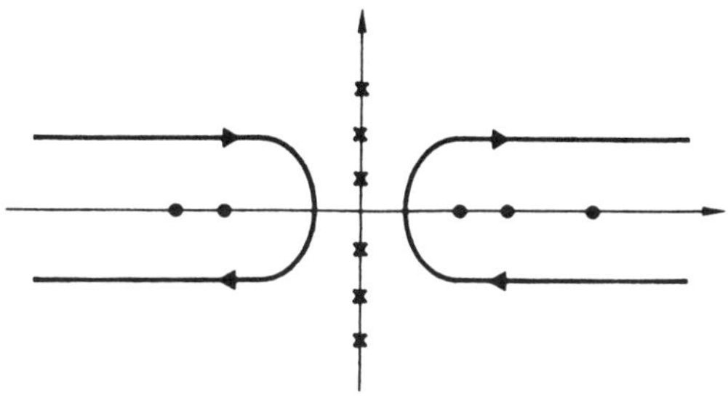

FIG. 10.4. Contours for the integrals (10.20).

shown in Fig. 10.4(b). As an example, let us compute the quark loop contribution to the partition function Z_Q, in the presence of time-independent fields. This is the first term of (4.20):

$$\ln Z_Q = Tr \ln (\partial_\tau + h) = \sum_{n\lambda} \ln (i\omega_n + e_\lambda) \tag{10.21}$$

We have evaluated the trace on the basis formed by the common eigenstates of ∂_τ and h:

$$\partial_\tau |\omega_n \lambda\rangle = i\omega_n |\omega_n \lambda\rangle \qquad h |\omega_n \lambda\rangle = e_\lambda |\omega_n \lambda\rangle \tag{10.22}$$

and we have chosen the fermion Matsubara frequencies (10.14) for ω_n. We apply (10.20) and we use the relation $\left(e^{-\beta z} + 1\right)^{-1} = \frac{1}{\beta}\frac{d}{dz} \ln \left(1 + e^{\beta z}\right)$. We deform the contour as in Fig. 10.4(b) and we integrate by parts. The result is the familiar expression for the partition function of a system of independent fermions, described by a hamiltonian h:

$$\ln Z_Q = Tr \ln (\partial_\tau + h) = \sum_{\lambda} \ln \left(1 + e^{-\beta e_\lambda}\right) \tag{10.23}$$

At finite density and temperature, a chemical potential μ simply changes the e_λ to $e_\lambda - \mu$. The possibility of deforming the integration contour shown in Fig. 10.4, may be jeopardized by the presence of regulators which are introduced to regularize the quark loop term. This would occur in the case of proper-time regularization or in the case of non-local regulators (see chapter 6). Indeed, the regulators can produce poles in the complex k_0 plane [85]. As long as only one-loop contributions are taken into account, the problem can be avoided by separating the action into a zero temperature part, which is regularized, and an unregularized thermal part, which is the difference between the action at finite temperature and zero temperature. For example, the actions (4.20) or(4.23) would be replaced by:

$$I_\beta (S,P,V,A) = I_{\Lambda,\beta\to\infty} (S,P,V,A) + [I_\beta (S,P,V,A) - I_{\beta\to\infty} (S,P,V,A)] \tag{10.24}$$

where $I_{\Lambda,\beta\to\infty} (S,P,V,A)$ is the regularized zero temperature action. The second term of (10.24) is the thermal contribution. For time-independent fields, the thermal contribution to the effective action is given by (10.23):

$$\begin{aligned} I_\beta (S,P,...) - I_{\beta\to\infty} (S,P,...) &= -Tr_\beta \ln (\partial_\tau + h) + Tr_{\beta\to\infty} \ln (\partial_\tau + h) \\ &= -\sum_{e_\lambda>0} \ln \left(1 + e^{-\beta e_\lambda}\right) + \sum_{e_\lambda<0} \ln \left(1 + e^{\beta e_\lambda}\right) \end{aligned} \tag{10.25}$$

It is free from ultraviolet divergences and it is not regularized, so that the deformed contour shown in Fig. 10.4(b) may be applied to it. Such a procedure, admittedly heuristic, can only be justified for temperatures $1/\beta$ which are well below the cut-off Λ. The second term of (10.24) however, is not, free of divergences when higher order loops are taken into account.

10.5.2 *The physical content*

Since the pions are lightest hadrons ($m_\pi = 139$ MeV) , there exists a temperature range $T \approx 50 - 100$ MeV at which only pions are excited. The probability of exciting other hadrons of mass $m > m_\pi$ is cut down by Boltzmann factors $e^{-\beta m}$, where $\beta = 1/T$ is the inverse temperature. The partition function reduces then to the partition function of non-interacting pions:

$$\ln Z_\pi = -\nu \sum_{\boldsymbol{k}} \ln \left(1 - e^{-\beta \varepsilon_k}\right) \qquad \varepsilon_k = \sqrt{k^2 + m_\pi^2} \tag{10.26}$$

where $\nu = 3$ is the isospin degeneracy of the pions. At low temperatures, the interactions between pions are suppressed [140]. In the chiral limit $m \to 0$, we can seek an expansion of the partition function in powers of $\beta m_\pi = \frac{m_\pi}{T}$ [141]:

$$\ln Z_\pi = \Omega \beta \nu T^4 \left(\frac{\pi^2}{90} - \frac{m_\pi^2}{24T^2} + \frac{m_\pi^3}{12\pi T^3} + ... \right) \tag{10.27}$$

High and low temperature expansions of thermodynamic quantities can be found in [142] for relativistic boson systems and in [143] for relativistic fermion systems.

The expansion (10.27) can be used to deduce a model-independent evaluation of the temperature dependence of the quark condensate. Since the current quark mass m appears in the QCD hamiltonian in the form $\int d_3 r \bar{q} m q$, the quark condensate at temperature $T = 1/\beta$ is given by:

$$\langle \bar{q}q \rangle_T = -\frac{1}{\beta \Omega} \frac{\delta \ln Z}{\delta m} \tag{10.28}$$

Let $\langle \bar{q}q \rangle_0$ be the vacuum quark condensate at zero temperature. The quark condensate $\langle \bar{q}q \rangle$ at finite temperature can be deduced from (10.27):

$$\langle \bar{q}q \rangle_T = \langle \bar{q}q \rangle_0 - \frac{1}{\beta \Omega} \frac{\delta \ln Z_\pi}{\delta m} = \langle \bar{q}q \rangle_0 + \frac{\delta}{\delta m} \nu T^4 \left(\frac{m_\pi^2}{24T^2} - \frac{m_\pi^3}{12\pi T^3} + ... \right) \tag{10.29}$$

From the Gell-Mann Oakes Renner relation $m \langle \bar{q}q \rangle_0 = -m_\pi^2 f_\pi^2$ we deduce that $\frac{\delta m_\pi}{\delta m} = -\frac{\langle \bar{q}q \rangle_0}{2m_\pi f_\pi^2}$, so that:

$$\langle \bar{q}q \rangle_T = \langle \bar{q}q \rangle_0 \left(1 - \frac{T^2}{8f_\pi^2} + ... \right) \tag{10.30}$$

The temperature dependence (10.30) of the quark condensate is a $1/N_c$ effect because f_π^2 is proportional to N_c. Suppose that we calculate the partition function in the Nambu Jona-Lasinio model. In section 5.3, we saw that, to leading order in N_c, the partition function of the system is given by the euclidean action: $\ln Z = -I(S, P, V, A)$. Let us separate the action into vacuum and thermal parts as in (10.24). The quarks are in plane wave orbits with energies $\pm \varepsilon_k$ where $\varepsilon_k = \sqrt{k^2 + M^2}$. The contribution of each orbit to the quark condensate is $\pm \frac{1}{\Omega} \frac{\delta \varepsilon_k}{\delta M} = \pm \frac{M}{\Omega \varepsilon_k}$. The change in the quark condensate at low temperatures can be deduced from (10.23) and (10.28):

$$\langle \bar{q}q \rangle_T - \langle \bar{q}q \rangle_{T=0}$$

$$= \frac{4N_c}{\Omega} \sum_{\boldsymbol{k}} \frac{2M}{\varepsilon_k} \frac{1}{e^{\beta \varepsilon_k} + 1} = \left(\frac{4N_c}{\pi^2} M^3 \right) \int_0^\infty \frac{x^2}{\sqrt{x^2+1}+1} e^{-\frac{M}{T}\sqrt{x^2+1}} \tag{10.31}$$

To leading order in M/T this expression can be reduced to [143]:

$$\langle \bar{q}q \rangle_T - \langle \bar{q}q \rangle_{T=0} \approx \frac{4N_c}{\pi^2} M^3 \int_1^\infty y dy \sqrt{1 - y^2} e^{-\beta M y} \approx \left(\frac{2MT}{\pi} \right)^{3/2} e^{-\frac{M}{T}} \tag{10.32}$$

where we used Eqns (9.6.23) and (9.7.2) of [144]. Table 5.1 shows that a constituent quark mass $M = 320$ MeV yields a quark condensate $\langle \bar{q}q \rangle^{1/3} = -276$ MeV. The expression (10.32) predicts a 1% reduction of the quark condensate

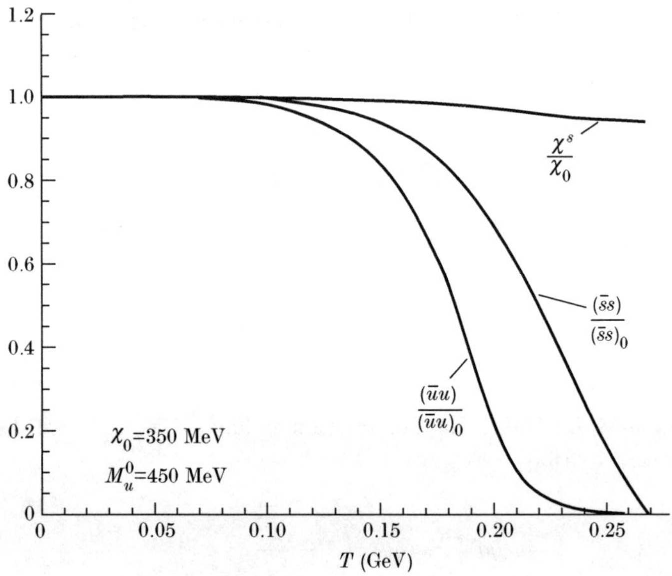

FIG. 10.5. Quark and gluon condensates are plotted as a function of temperature. The case of a high gluon condensate $\chi_0 = 350$ MeV, where the gluon condensate does not feel the chiral phase transition. From [55].

at the temperature $T \doteq 100$ MeV. This is the quark loop contribution, which is leading order in N_c. The estimate (10.30), which includes thermal excitation of pions, predicts a 14% reduction of the quark condensate. This is an instance where the $1/N_c$ expansion is not valid. The reason is quite simple. The quark loop contribution takes into account thermal excitations of quarks with mass $M = 320$ MeV. The probability of such excitations is reduced by Boltzmann factors $e^{-M/T}$. At low temperatures, the thermal excitations are dominated by the massless pions. If the Nambu Jona-Lasinio model is pushed to next to leading order in N_c, which includes the quantum fluctuations of the fields S and P, then the model-independent estimate (10.30) is obtained correctly [105], as it should be, since the pions are the quantum fluctuations of the P field. The temperature dependence of next to leading order effects is also studied in [106, 107].

10.5.3 *Restoration of chiral symmetry at high temperatures*

Pions no longer dominate when the temperature becomes comparable to the constituent quark mass M. An appreciable fraction of Dirac sea quarks are then thermally excited to the positive energy orbits. They reduce the constituent quark mass, until eventually chiral symmetry is restored. Restoration of chiral symmetry may be studied in the large N_c limit. A reduction of about 10% of the transition temperature is estimated to result from the next to leading order contribution [105]. At finite temperature, the effective action is modified by the

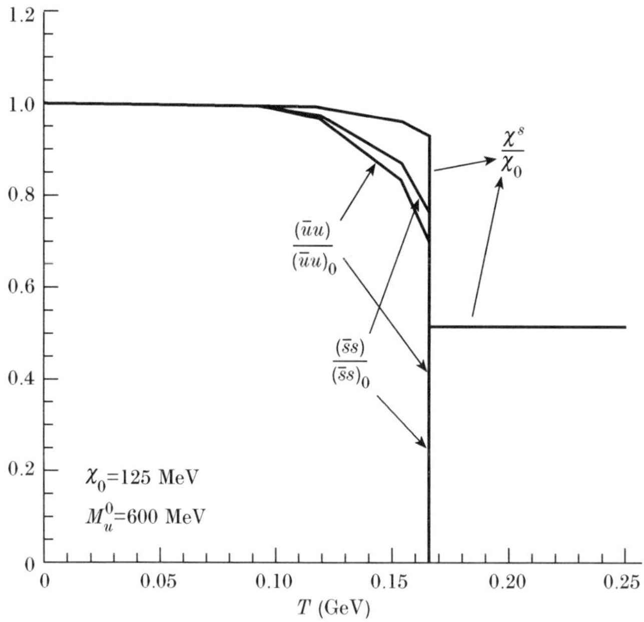

FIG. 10.6. Quark and gluon condensates are plotted as a function of temperature. A low value $\chi_0 = 125$ MeV of the gluon condensate is assumed and the gluon condensate follows the chiral phase transition. From [55].

thermal contribution, which is the second term of (10.24). The gap equation is obtained by searching for a stationary point of the action, as in the zero temperature case. Quark condensates can be obtained from (10.28).

Figures 10.5 and 10.6, taken from [55], show two examples of quark and gluon condensates plotted against the temperature. They are calculated from a scaled Nambu Jona-Lasinio action (section 11.2) of the form (11.4) with $N_f = 3$ flavors. Figure 10.5 shows the case of a high vacuum gluon condensate $\chi_0 = 350$ MeV. The gluon condensate is only weakly coupled to the quark condensate and it is not significantly modified by the chiral phase transition. An unscaled Nambu Jona-Lasinio action would yield a similar curve. The transition temperature is close to 190 MeV. Figure 10.6 shows the case of a lower vacuum gluon condensate $\chi_0 = 125$ MeV. In this case the quark and gluon condensates are strongly coupled. The chiral phase transition is abrupt and felt by the gluon condensate. Notice also that the transition temperature is close to 160 MeV, somewhat lower than in the case depicted on Fig.10.5. In fact, unscaled Nambu Jona-Lasinio models are unable to yield transition temperatures lower than about 180 MeV. In scaled models (section 11.2), transition temperatures as low as $T_c \approx 140$ MeV can only be obtained by adopting low values of the gluon condensate [145]. The problem is that chiral symmetry is then be restored at densities close to or even lower than normal nuclear matter density. In order to prevent this, it may prove necessary to introduce the a_1 field, which has the property of stiffening the vacuum against the restoration of chiral symmetry [30, 132]. Further applications to systems at finite temperature are reviewed in [36, 52, 106, 107, 137, 146, 147, 148, 149, 139, 150, 151].

11

FURTHER CHIRAL QUARK MODELS

The Nambu Jona-Lasinio is but one of several models which describe quarks interacting with chiral fields. In this chapter we review and compare other models which have been extensively used. They all have in common a quark loop term which describes quarks interacting with a chiral field.

11.1 The Diakonov Petrov model.

The Diakonov Petrov model [152] is defined by the action:

$$I(q, \bar{q}, \theta) = \int d_4 x \bar{q} \left(-i\partial_\mu \gamma_\mu + m + M e^{i\gamma_5 \theta_a \tau_a}\right) q \tag{11.1}$$

where the τ_a are the $N_f^2 - 1$ matrices of the $SU(N_f)$ group defined in appendix A.1 and where m is a current quark mass, assumed to have the diagonal form (4.2) in flavor space. The mass M plays the role of a constituent quark mass. The quark field can be integrated out and the quark loop regularized:

$$I(\theta) = -Tr_\Lambda \ln\left(-i\partial_\mu \gamma_\mu + m + MU\right) \qquad U = e^{i\gamma_5 \theta_a \tau_a} \tag{11.2}$$

In principle, M is a momentum-dependent operator, of the form (3.11), which can regularize the theory. However, most applications of the Diakonov Petrov model have been made with a proper-time regularized action. The $\theta_a(x)$ are the dynamical variables of the system which, in the chiral limit $m \to 0$, describe $N_f^2 - 1$ Goldstone bosons.

This model is deduced from a bona fide calculation of the QCD vacuum, which takes the form of an instanton liquid. Chiral symmetry is shown to be spontaneously broken by quarks propagating in the instanton liquid [26, 27, 31]. It is argued that, at energies smaller than $\frac{1}{\rho} \approx 600$ MeV, where ρ is the average size of the instantons, the only degrees of freedom are the fields describing the Goldstone bosons *and* the quarks. All other degrees of freedom are frozen out. The cut-off is also of the order of $\frac{1}{\rho}$, as shown in the form (3.11).

In the $SU(2)_f$ case, the chiral field can be written in the form:

$$MU = M e^{i\gamma_5 \theta_a \tau_a} = S + i\gamma_5 P_a \tau_a \qquad S^2 + P_a^2 = M^2 \tag{11.3}$$

The model restricts the scalar and pseudoscalar fields S and P to remain on the chiral circle, as in the non-linear sigma model. This is essentially what distinguishes the Diakonov Petrov model from the Nambu Jona-Lasinio model, described by the action (5.24). The Diakonov Petrov model has been extensively used for soliton calculations (chapter 12). One compelling reason is that the Nambu Jona-Lasinio model does not lead to a stable soliton, unless the chiral field is limited to the chiral circle, as in (11.3), in which case it reduces to the Diakonov Petrov model (in the absence of vector interactions) [153, 154]. Variants of the Diakonov Petrov model have been studied, in which extra terms are added which give the quarks an axial coupling constant $g_{A(quark)} \neq 1$. The linear relation between the quark and nucleon axial coupling constants, speculated in [155] , is not borne out in soliton calculations [156]. In section 13.5 it is shown that the observed nucleon axial coupling constant g_A can be obtained with $g_{A(quark)} = 1$ when the effect of the soliton rotation is taken into account.

11.2 Scaled models involving a dilaton field

Most models do not implement the QCD scale or trace anomaly [157, 158]. This means that their action is not scale-invariant, as the QCD action is, in the limit of vanishing quark masses. Consider, for example, the Nambu Jona-Lasinio lagrangian (4.1). In a scale transformation, the quark field transforms as $q(x) \rightarrow \lambda^{3/2} q(\lambda x)$, meaning that it has a dimension 3/2. The four-fermion interactions have dimension 6, so that they do not lead to a scale-invariant action $\int d_4 x \mathcal{L}$.

Scale invariance of a model lagrangian can be restored by introducing an extra degree of freedom, in the form of a scalar *dilaton field* $\chi(x)$, which transforms with a dimension 1: $\chi(x) \rightarrow \lambda \chi(\lambda x)$. For example, if the coupling constants g^2 of the Nambu Jona-Lasinio model are divided by the field χ^2, then the four-fermion interactions acquire dimension 4 and the action $\int d_4 x \mathcal{L}$ becomes scale-invariant. In the models discussed below, the dilaton field is treated as a dynamical variable.

Let us illustrate the method for the Nambu Jona-Lasinio action (5.24) in the $N_f = 2$ flavor case. It can easily be extended to include vector mesons [159] as well as to $N_f = 3$ flavors [55]. One way to make the action scale-invariant is to write it in the form:

$$I(S,P,\chi) = -Tr_{\chi\lambda}\left(-i\partial_\mu\gamma_\mu + m + S + i\gamma_5 P_a \tau_a\right) + \frac{\chi^2}{2g_s^2}\int d_4 x \left(S^2 + P_a^2\right)$$

$$+ \int d_4 x \left(\frac{1}{2}\left(\partial_\mu \chi\right)^2 + \frac{b^2}{16}\left(\chi^4 \ln\frac{\chi^4}{\chi_G^4} - \left(\chi^4 - \chi_G^4\right)\right)\right) \tag{11.4}$$

In the first line, the dilaton field χ has been introduced so as to make the action scale-invariant (in the chiral limit $m \rightarrow 0$). It is not sufficient to divide the coupling constant g_s^2 by χ^2. The cut-off Λ also needs to be proportional to χ so that it would scale with dimension 1. The cut-off has therefore been written in the form $\Lambda = \chi\lambda$. The chiral field $S + i\gamma_5 P$ scales with dimension 1.

The second line of (11.4) is the action of the field χ. It is considered as a dynamical field with a kinetic term $\frac{1}{2}\left(\partial_\mu \chi\right)^2$. The potential term, which is proportional to a dimensionless constant b^2, has a minimum at $\chi = \chi_G$. It explicitly breaks scale invariance. Its form is chosen such that the scale invariance is broken in the same way as in the scale anomaly of QCD [160]. Namely, if one calculates the Noether current s_μ, associated to the scale transformation of the fields, then one finds that $\partial_\mu s^\mu = \theta^\mu_\mu = \frac{b^2}{4}\chi^4$, where $\theta^{\mu\nu}$ is the energy momentum tensor. The QCD scale anomaly reads $\theta^\mu_\mu = \frac{\beta_{QCD}}{2g}\left\langle F^a_{\mu\nu}F^{\mu\nu}_a\right\rangle$. This leads us to interpret the dilaton field as representing the *gluon condensate*: $\chi^4 \sim \left\langle F^a_{\mu\nu}F^{\mu\nu}_a\right\rangle$. Whether one should identify the glueball which is to be associated to the dilaton field χ with one of the $f_0\,(1350)$, $f_0\,(1525)$ or $f_0\,(1590)$ resonances is still a subject of debate [66].

In section 5.4 we derived an equivalent sigma model (5.32) to the Nambu Jona-Lasinio action (5.24). The same procedure can be used to derive an equivalent sigma model to the scaled action (11.4), with the result [161]:

$$\mathcal{L}_{\sigma-model}(\sigma,\pi) = \frac{1}{2}\left(\partial_\mu\sigma\right)^2 + \frac{1}{2}\left(\partial_\mu\pi_a\right)^2 + \frac{m_\pi^2}{2}\pi_a^2 + \frac{m_\sigma^2}{8\sigma_0^2}\left(\sigma^2 - \frac{\sigma_0^2}{\chi_0^2}\chi^2\right)^2$$

$$+\frac{1}{2}\chi^2 + \frac{m_{GL}^2}{8\chi_0^2}\left(\chi^2 - \chi_0^2\right)^2 \qquad (11.5)$$

The σ and π fields are given by (5.30). The vacuum value of the χ field is χ_0. It is shifted from the minimum χ_G of the dilaton potential, because of the coupling of the χ field to the scalar field S. The masses m_σ and m_{GL} are given by the expressions:

$$m_\sigma^2 = 4M^2 \qquad m_{GL}^2 = \chi_0^2\left(b^2 - \frac{N_c\Lambda^4}{\pi^2}\right) \qquad (11.6)$$

where M is the vacuum value of the scalar field S. In the case of $N_f = 3$ flavors, the second term of m_{GL}^2 is multiplied by a factor of 3/2 [55]. The value of m_{GL} is usually taken to be close to 1.5 GeV where a glueball is expected to be found. The fourth term of (11.5) is a coupling between the scalar fields σ and χ. This is effectively a coupling between the quark and gluon condensates, because $\sigma \sim S \sim \langle\bar{q}q\rangle$ and $\chi^4 \sim \left\langle F^a_{\mu\nu}F^{\mu\nu}_a\right\rangle$. The masses of the physical mesons can only be identified after diagonalizing the mass matrix of the equivalent sigma model (11.5).

Such scaled models can be used to study the variation of quark and gluon condensates in dense and hot systems [55, 139, 159, 161, 162]. An example is given in Fig.11.1, where the quark and gluon condensates are shown as a function of the baryonic density.

Model-independent estimates of the variation of the gluon condensate with density [129] suggest that, at normal nuclear density, the gluon condensate does not decrease more than about 5%. In a scaled Nambu Jona-Lasinio model, the variation of the gluon condensate depends very much on the chosen vacuum

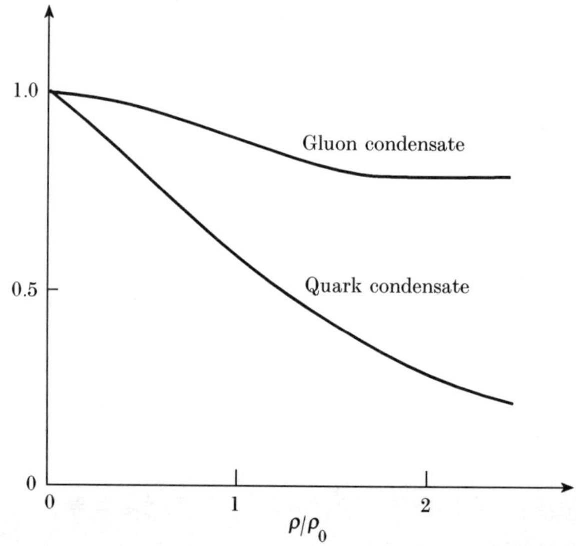

FIG. 11.1. Quark and gluon condensates are plotted as a function of density. From [159].

value χ_0 of the gluon condensate and on the constituent quark mass M. Indeed, the dilaton field feels the baryonic density, carried by the quarks, only inasmuch as it couples to the sigma field, through the term $\frac{M^2}{\chi_0^2}\sigma^2\chi^2$. When one uses the value $\chi_0 \approx 350$ MeV for the gluon condensate, which is deduced from QCD sum rules applied to charmonium [43, 44], then the coupling of the quark and gluon condensates is weak and the dilaton field is frozen to its vacuum value. The scaled action (11.4) does not then differ significantly from the action of the unscaled Nambu Jona-Lasinio model. However, lower values $\chi_0 \approx 100$ MeV have been used in order to enhance the coupling. Such values however, often induce a chiral phase transition at densities equal to or lower than normal nuclear matter density. Low values of the gluon condensate are required to obtain a chiral phase transition at temperatures as low as $T \approx 140\ MeV$. High values of the constituent quark mass are required to reproduce the lattice estimates of the entropy below the phase transition [145, 163]. The coupling of the pion and a_1 fields could help however, by stiffening the vacuum against the restoration of chiral symmetry [30].

The nature of the chiral phase transition (critical density, order of the phase transition) depends on the model parameters. It also depends on the form chosen to implement the QCD scale anomaly. The latter is far from being unique and a variety of forms have been studied [55, 159], which yield very different variations of condensates and meson masses as a function of density and temperature. Instead of using the action (11.4), we could equally well have used the action:

$$I(S,P,\chi) = -Tr_{\chi\lambda}(-i\partial_\mu\gamma_\mu + m + S + i\gamma_5 P_a\tau_a) + \frac{1}{2g_s^2}\int d_4x\left(S^2+P_a^2\right)^2 + I(\chi) \tag{11.7}$$

This model differs from the model described by the action (11.4) in that the chiral field appears as a fourth order polynomial, which is scale-invariant and which therefore does not need to couple to the dilaton field. With the model (11.7) the scalar field $S \approx \langle \bar{q}q \rangle^{1/3}$ whereas with the model (11.4) $S \approx \langle \bar{q}q \rangle$. As a result, the model (11.7) predicts meson masses which decrease with density in proportion to $\langle \bar{q}q \rangle^{1/3}$, as once suggested [164, 165]. The model (11.7) also yields a σ meson mass about twice as large as $2M$. The variety of behavior displayed by models in which actions are scaled in different ways shows that symmetry is far from sufficient to determine the behavior of a hadronic system.

The procedure of implementing the scale anomaly of QCD has been used in connection with chiral quark models [160, 166] as well as with Nambu Jona-Lasinio models [55, 159, 161, 167] . The dilaton field has also been introduced into Skyrme models to act as the chiral partner of the pion [164, 165]. Indeed, the Weinberg lagrangian (10.1) can be made scale-invariant as follows:

$$\mathcal{L} = \frac{f_\pi^2 \chi^2}{4\chi_G^2} tr_f \left(\partial_\mu U\right)\left(\partial_\mu U^\dagger\right) + \frac{c\chi^3}{4\chi_G^3} tr_f m \left(U + U^\dagger - 2\right) \tag{11.8}$$

$$+\frac{1}{2}\left(\partial_\mu \chi\right)^2 + \frac{b^2}{16}\left(\chi^4 \ln \frac{\chi^4}{\chi_G^4} - \left(\chi^4 - \chi_G^4\right)\right) \tag{11.9}$$

It becomes then equivalent to a linear sigma model. In $SU(2)_f$ we would have $\chi U = S + iP_a\tau_a$ and $\chi^2 = S^2 + P_a^2$. A scale-invariant Skyrme-like 4-derivative term could be added. The current quark mass term $\frac{\chi^3}{\chi_G^3}$ is scaled so as to give it a dimension 3 as in QCD where the current quark mass term is $\bar{q}mq$. On the assumption that the constant c is proportional the quark condensate $\langle \bar{q}q \rangle$, the effective lagrangian (11.8) has also been used to suggest that hadron masses should remain proportional to $\langle \bar{q}q \rangle^{1/3}$ in systems at finite density [164, 165].

Scaled models have quite distinct properties from the Nambu Jona-Lasinio model [143]. With $N_f = 3$ flavors, the $\langle \bar{u}u + \bar{d}d \rangle$ and $\langle \bar{s}s \rangle$ condensates become coupled, so that both the $\langle \bar{u}u + \bar{d}d \rangle$ and $\langle \bar{s}s \rangle$ condensates undergo the same chiral phase transition. They give rise to a discontinuous first order phase transition, whereas the unscaled Nambu Jona-Lasinio model always gives rise to a second order transition [52, 147]. They also allow hadron masses to scale in proportion to $\langle \bar{q}q \rangle^{1/3}$ [159].

11.3 The Gell-Mann Lévy model

The Gell-Mann Lévy model [168] is a linear sigma model with, in addition, quarks coupled to the chiral field. It is the first chiral quark model, with which solitons were calculated [169, 170, 171]. A useful review can be found in [172]. In this section, we focus on the differences between the Gell-Mann Lévy model and the Nambu Jona-Lasinio model.

11.3.1 *The action of the model*

The model is defined by the lagrangian density [173]:

$$\mathcal{L} = \bar{q}\left(i\partial_\mu\gamma^\mu - m - g\left(\sigma + i\gamma_5\pi\right)\right)q$$

$$+\frac{1}{2}\left(\partial_\mu\sigma_a\right)\left(\partial^\mu\sigma_a\right) + \frac{1}{2}\left(\partial_\mu\pi_a\right)\left(\partial^\mu\pi_a\right) - \frac{\lambda^2}{8}\left[\sigma_a^2 + \pi_a^2 - f^2\right]^2$$

$$-\frac{\eta^2}{4}\left[\frac{1}{3}tr_f\left[\left(\sigma + i\pi\right)\left(\sigma - i\pi\right)\right]^2 - \left[\frac{1}{3}tr_f\left(\sigma + i\pi\right)\left(\sigma - i\pi\right)\right]^2\right]$$

$$+\frac{\kappa^2}{2}\left(\det\left(\sigma + i\pi\right) + \det\left(\sigma - i\pi\right)\right) \tag{11.10}$$

where $\sigma = \sigma_a\tau_a$ and $\pi = \pi_a\tau_a$ are the scalar and pseudoscalar meson fields written in the form of the hermitian $N_f \times N_f$ matrices (5.48) and m is a current quark mass matrix of the form (4.2). The τ_a are the N_f^2 generators of the $U(N_f)$ group (section A.2) with $\tau_{a=0} = \sqrt{\frac{2}{N_f}}I$. The coupling constants g, λ and η are dimensionless. The last term of (11.10) is a $U_A(1)$ symmetry breaking term of the form (4.35). The Gell-Mann Lévy model (11.10) has the same symmetries as the Nambu Jona-Lasinio model (see section 4.5). In the chiral limit $m \to 0$, it is invariant in chiral transformations (4.30) and it gives rise to massless Goldstone bosons.

In the simplified case of $N_f = 2$ flavors, we have $\left(\sigma + i\pi\right)\left(\sigma - i\pi\right) = \sigma_a^2 + \pi_a^2$, with $a = 1, 2, 3$, so that the term proportional to η^2 vanishes. If we further simplify by retaining only one scalar field and 3 pseudoscalar fields, the lagrangian (11.10) becomes equivalent to the original lagrangian of Gell-Mann and Lévy [168]:

$$\mathcal{L} = \bar{q}\left(i\partial_\mu\gamma^\mu - m - g\left(\sigma + i\gamma_5\pi_a\tau_a\right)\right)q$$

$$+\frac{1}{2}\left(\partial_\mu\sigma\right)\left(\partial^\mu\sigma\right) + \frac{1}{2}\left(\partial_\mu\pi_a\right)\left(\partial^\mu\pi_a\right) - \frac{\lambda^2}{8}\left[\sigma + \pi_a^2 - f_\pi^2\right]^2 \tag{11.11}$$

Applications of the $SU(3)_f$ Gell-Mann Lévy model are discussed in [173]. For the sake of clarity, we shall discuss the simplified Gell-Mann Lévy model (11.11). The arguments can readily be extended to the more complete $SU(3)_f$ case (11.10).

11.3.2 *Comparison of the Gell-Mann Lévy and the Nambu Jona-Lasinio models*

We may proceed as in sections 4.2 and 4.3. We derive a path integral for the partition function, of the form (4.23). It is expressed in terms of the euclidean action:

$$I\left(\sigma, \pi\right) = -Tr\ln\left(-i\partial_\mu\gamma_\mu + m + g\left(\sigma + i\gamma_5\pi_a\tau_a\right)\right)$$

$$+\int d_4x\ \left[\frac{1}{2}\left(\partial_\mu\sigma\right)^2+\frac{1}{2}\left(\partial_\mu\pi_a\right)^2+\frac{\lambda^2}{8}\left[\sigma^2+\pi_a^2-f_\pi^2\right]^2\right] \tag{11.12}$$

We can express the action in terms of the redefined fields:

$$S=g\sigma \qquad M=gf_\pi \qquad P_a=g\pi_a \tag{11.13}$$

The action (11.12) becomes:

$$I\left(S,P\right)=-Tr\ln\left(-i\partial_\mu\gamma_\mu+m+S+i\gamma_5P_a\tau_a\right)$$
$$+\frac{1}{2g^2}\int d_4x\ \left(\left(\partial_\mu S\right)^2+\left(\partial_\mu P_a\right)^2\right)+\frac{\lambda^2}{8g^4}\int d_4x\ \left(S^2+P_a^2-M^2\right)^2 \tag{11.14}$$

Let us compare the Gell-Mann Lévy action (11.14) with the Nambu Jona-Lasinio action (5.24). They describe similar physics although the fields have quite different meaning.

Formally, the main difference between the two actions is the presence of kinetic terms for the fields S and P, represented by the second term of (11.14). In the Gell-Mann Lévy model, the meson fields σ and π are dynamical variables that exist even in the absence of the quarks. In the Nambu Jona-Lasinio model, the fields $S=-g_s^2\bar{q}q$ and $P_a=-g_s^2\bar{q}i\gamma_5\tau_aq$ are composite $q\bar{q}$ excitations. In the Gell-Mann Lévy model, the fields are only partly composite $q\bar{q}$ excitations and chiral symmetry is broken even in the absence of quarks.

Another major, although related, difference is the renormalizability of the Gell-Mann Lévy model. To renormalize the action at the quark loop level, we expand the quark loop term about the point $S=M$, $P=0$. To simplify the argument further, we consider the chiral limit $m\to 0$. We redefine the action (11.14) by subtracting its value at the point $S=M$, $P=0$ and we write it in the form (see section 5.1):

$$I\left(S,P\right)=-\frac{1}{2}Tr\ln\left(1+GV\right)$$
$$+\frac{1}{2g^2}\int d_4x\ \left(\left(\partial_\mu S\right)^2+\left(\partial_\mu P_a\right)^2\right)+\frac{\lambda^2}{8g^4}\int d_4x\,,\left(S^2+P_a^2-M^2\right)^2 \tag{11.15}$$

where G and V are defined in (5.9) and (5.10). We renormalize the action by subtracting the infinite part of the quark loop term. This is justified because, as shown in section 5.2, the divergent part of the quark loop term $-\frac{1}{2}Tr\ln\left(1+GV\right)$ is:

$$-\frac{1}{2}TrGV+\frac{1}{4}TrG^2V^2=-a\int d_4x\left(S^2+P_a^2-M^2\right)$$
$$+b\int d_4x\left(\left(\partial_\mu S\right)^2+\left(\partial_\mu P_a\right)^2+\left(S^2+P_a^2-M^2\right)^2\right) \tag{11.16}$$

where a and b are (infinite) constants (see equation (5.27) for example). The subtraction of the infinite part (11.16) of the quark loop term is equivalent to a redefinition (renormalization) of the coupling constants g and λ. (Strictly, we should have added to the Gell-Mann Lévy action (11.14) a term of the form $\frac{\mu^2}{2g^2}\left(S^2+P_a^2-M^2\right)$ which we would have subsequently set to zero, after renormalizing μ.) The finite renormalized Gell-Mann Lévy action is thus:

$$I_R(S,P)=-\frac{1}{2}Tr\ln(1+GV)+\frac{1}{2}TrGV-\frac{1}{4}TrG^2V^2$$

$$+\frac{1}{2g^2}\int d_4x\ \left((\partial_\mu S)^2+(\partial_\mu P_a)^2\right)+\frac{\lambda^2}{8g^4}\int d_4x\ \left(S^2+P_a^2-M^2\right)^2 \tag{11.17}$$

The counter-terms in (11.17) are chosen such that the quark loop makes no contribution to the first and second derivatives of the effective potential $\frac{\lambda^2}{8g^4}\ \left(S^2+P_a^2 -M^2\right)^2$ with respect to $\left(S^2+P_a^2\right)$. This defines the renormalization conditions. The point $S=M,\ P=0$ remains a stationary point of the action. We notice that the counter-term $-\frac{1}{4}TrG^2V^2$ subtracts, from the Gell-Mann Lévy action, *the full contribution of the quark loop to the equivalent linear sigma model* (5.20) *of the Nambu Jona-Lasinio action.* In other words, the main part of the quark loop contribution of the Nambu Jona-Lasinio model is absorbed into the meson terms of the Gell-Mann Lévy model. This substantially reduces the contribution of the quark loop to the renormalized Gell-Mann Lévy model.

11.3.3 *The Landau ghost*

We can calculate the inverse correlation function of the scalar field S by expanding the renormalized action (11.17) to second order in the fluctuating parts $\widetilde{S}=S-M$ of the field. Proceeding as in section 9.2, we find the following expression for the inverse correlation function:

$$K_{\widetilde{S}}^{-1}(q)=\frac{1}{g^2}\left(q^2+m_\sigma^2+4N_cg^2\left(q^2+4M^2\right)f_R(q)\right) \tag{11.18}$$

where the renormalized loop integral $f_R(q)$ is:

$$f_R(q)=\frac{1}{\beta\Omega}\sum_q$$

$$\left(\frac{1}{\left(\left(k-\frac{q}{2}\right)^2+M^2\right)\left(\left(k+\frac{q}{2}\right)^2+M^2\right)}-\frac{1}{\left(k^2+M^2\right)\left(k^2+M^2\right)}\right) \tag{11.19}$$

and where the σ meson mass is $m_\sigma=2\lambda f_\pi=\frac{2\lambda M}{g}=\frac{2\lambda f_\pi}{g^2}$. The loop integral (11.19) can be evaluated analytically:

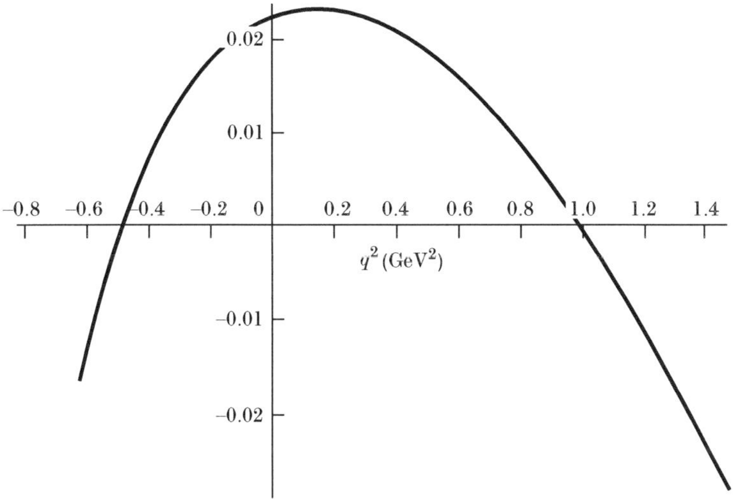

FIG. 11.2. The inverse correlation function (11.18) of the scalar field is plotted as a function of q^2. The zero for a negative q^2 corresponds to an on-shell meson. The zero for a positive q^2 is the Landau ghost.

$$f_R(q) = -\frac{\pi^2}{(2\pi)^4}\left(-2 + 2\frac{\sqrt{q^2+4M^2}}{q} arctanh \frac{q}{\sqrt{q^2+4M^2}}\right) \qquad q^2 > 0$$

$$f_R(q) = -\frac{\pi^2}{(2\pi)^4}\left(-2 + 2\frac{\sqrt{q^2+4M^2}}{q} arctan \frac{q}{\sqrt{q^2+4M^2}}\right) \qquad q^2 < 0 \tag{11.20}$$

Figure (11.2) shows a plot of the inverse correlation function (11.18) as a function of q^2. The correlation function is calculated with the following parameters:

$$m_\sigma = 600 \text{ MeV} \qquad M = 400 \text{ MeV} \qquad g = 4 \qquad N_c = 3 \tag{11.21}$$

In order for solitons to exist in the Gell-Mann Lévy model, the coupling constant needs to be $g \approx 4$ or larger. The constituent quark mass is then $M = gf_\pi \approx 400$ MeV or larger. This agrees with the constituent quark masses required for solitons to exist in the Nambu Jona-Lasinio model [174] (see section 12.8). The figure shows that the inverse correlation function has two zeros, which correspond to poles of the correlation function. One occurs on the negative q^2 axis at $q^2 \approx -0.5$ GeV^2. This pole corresponds to an on-shell particle with mass $m \approx 700$ MeV (see section 7.2.1). It is the mass of the σ meson, which is raised by 100

MeV by the quark loop contribution, relative to its input value $m_\sigma = 600$ MeV. The second zero occurs on the positive q^2 axis at $q^2 \approx 1$ GeV^2. This pole is a Landau ghost, which signals an instability of the vacuum [95, 96]. The meaning of Landau ghosts is discussed in section 7.2.2.

The position of the Landau pole can be estimated from the asymptotic form of the inverse scalar meson propagator (11.18), valid for large q:

$$K_S^{-1}(q) = \frac{q^2}{g^2}\left(1 - 4N_c g^2 \frac{\pi^2}{(2\pi)^4} \ln \frac{q^2}{M^2}\right) \tag{11.22}$$

The inverse propagator vanishes when

$$q^2 = M^2 e^{\frac{4\pi^2}{N_c g^2}} \tag{11.23}$$

In addition, the residue of the pole is negative. For the values (11.21), the exponential factor in (11.23) is $e^{\frac{4\pi^2}{N_c g^2}} \approx 2.3$. The Landau ghost occurs at momenta close to 1 GeV^2, an uncomfortably low value. For higher constituent quark masses, the Landau ghost occurs at lower values of q^2. One effect of the Landau ghost is to lower the energy of solitons indefinitely as the gradients of the chiral fields are increased [95], thereby preventing the minimization of the soliton energy with respect to the shape of the profile function (see chapter 12). This is one of the reasons why the Gell-Mann Lévy model is no longer used for soliton calculations. Attempts have been made to eliminate the Landau ghost [97, 175]. In quantum electrodynamics, a well-known Landau ghost occurs at $q^2 = m^2 e^{\frac{3\pi}{e^2}}$, [5] where m is the electron mass and $e^{\frac{3\pi}{e^2}} \approx 10^{556}$, an incomparably higher value of q^2 than in the strongly interacting Gell-Mann Lévy model.

11.4 Quark confinement in color dielectric models

Quark confinement has been described in terms of an effective QCD lagrangian, supplemented by a scalar field χ:

$$L = \bar{q}\left(i\partial_\mu \gamma^\mu - \frac{m}{\chi} + \frac{g}{\chi} A_\mu \gamma^\mu\right) q - \chi^4 F_{\mu\nu} F^{\mu\nu} + \frac{1}{2}(\partial_\mu \chi)(\partial^\mu \chi) - V(\chi) \tag{11.24}$$

where A_μ is the gluon field and $F_{\mu\nu}$ the corresponding field tensor. The field χ acts as a color dielectric constant $\varepsilon = \chi^4$ to the gluon field [176]. The potential $V(\chi)$ has a minimum at $\chi = 0$, which is the value which the field takes in the confined phase (the physical vacuum). In the perturbative phase, the field acquires the value $\chi = 1$, in which case the lagrangian (11.24) reduces to the QCD lagrangian. In applications, gluons are either ignored or treated perturbatively. Solitons have been calculated from the simplified lagrangian:

$$L = \bar{q}\left(i\partial_\mu \gamma^\mu - \frac{m}{g\chi}\right) q + \frac{1}{2}(\partial_\mu \chi)(\partial^\mu \chi) - V(\chi) \tag{11.25}$$

Localized soliton solutions have been found, inside which $0 < \chi < 1$ so that the quarks acquire a finite constituent mass [177]. Outside the soliton, the field χ

vanishes and the quarks acquire an infinite mass. The quarks are thus confined inside the soliton as in bag models. Solitons described by confining lagrangians, such as (11.25) are quite similar to bag models, reviewed in [178]. Chiral symmetry can be implemented by replacing the quark mass term by a chiral field: $m \to \sigma + i\gamma_5 \pi_a \tau_a$. However, when this is done, the chiral field does not acquire a twist sufficient to give it a winding number [179, 180] and the pion field can be treated perturbatively as in the cloudy bag model. As an example, we write down how the color dielectric model (11.25) has been extended to $SU(3)_f$ [181]:

$$
\begin{aligned}
\mathcal{L} = \bar{q}\left(i\partial_\mu \gamma^\mu - m - \frac{g}{\chi}(\sigma + i\gamma_5\pi)\right) q + \frac{1}{2}(\partial_\mu \chi)(\partial^\mu \chi) - V(\chi) \\
+\frac{1}{2}(\partial_\mu \sigma_a)(\partial^\mu \sigma_a) + \frac{1}{2}(\partial_\mu \pi_a)(\partial^\mu \pi_a) - \frac{\lambda^2}{8}\left[\sigma_a^2 + \pi_a^2 - f^2\right]^2 \\
-\frac{\eta^2}{4}\left[\frac{1}{3} tr_f \left[(\sigma + i\pi)(\sigma - i\pi)\right]^2 - \left[\frac{1}{3} tr_f (\sigma + i\pi)(\sigma - i\pi)\right]^2\right] \\
+\frac{\kappa^2}{2}\left(\det(\sigma + i\pi) + \det(\sigma - i\pi)\right)
\end{aligned}
\tag{11.26}
$$

This lagrangian may be viewed as a Gell-Mann Lévy model (11.10), modified by the addition of the confining field χ. Color-dielectric models are reviewed in [172, 180]. Recent applications of the model (11.26) are discussed in [182].

Insurmountable divergences appear when we attempt to include Dirac sea quarks. For this reason, confining lagrangians, such as (11.25) and (11.26), have been limited to the description of valence quarks. At present we do not know how to use them for the description of vacuum dynamics.

12

CHIRAL SOLITONS

The possibility of describing baryons as chiral solitons is based on the observation that bound states can be formed when quarks interact with a chiral field. Friedberg and Lee had previously shown how bound states (called solitons) could be formed when quarks interact with scalar fields (see the reviews in [183, 184]). Chiral solitons were discovered when the scalar field was supplemented by a pion field, so as to implement chiral symmetry. This lead naturally to the calculation of solitons using the Gell-Mann Lévy model (section 11.3), also extended to include vector fields [169, 170, 171]. These, as well as solitons calculated with color dielectric models, are reviewed in [172]. Quite similar solitons are obtained using the Nambu Jona-Lasinio model. They are reviewed in [185, 186].

Solitons obtained with scalar and chiral fields are however very different. When quarks interact with scalar fields, as in the Friedberg–Lee model, the binding of the quarks can be understood in terms of a reduction of the constituent quark mass inside the soliton. The scalar meson field, which is proportional to the constituent quark mass, varies from f_π outside the soliton to almost zero inside. The variation of the scalar field produces a volume energy. The soliton attains an equilibrium size, in which the volume energy tries to shrink the soliton while the kinetic energy of the quarks tries to expand it, as in the MIT bag model [187, 188].

When quarks interact with a chiral field, as in the Gell-Mann Lévy model, the soliton could also acquire a volume energy by allowing the chiral field to deviate from the chiral circle. This is not however what is found to be energetically favoured. Instead, a bound state (soliton) with lower energy is formed when the chiral field acquires a hedgehog shape (see section 12.1), as in the Skyrmion (section 12.2). As one enters the soliton from the outside, the chiral field acquires a twist, called the winding number, while remaining close to the chiral circle, as in the Skyrme model. In such a case, no volume energy is generated and the constituent quark mass remains essentially constant.

The existence of a chiral field inside the nucleon and the existence of a volume energy remain open questions. Some recent calculations using constituent quark models with confining forces show that it is at least as accurate to assume chirally symmetric pion exchange between quarks in the nucleon as the usually assumed gluon exchange [4, 28]. In the Gell-Mann Lévy model, the twist of the chiral field is energetically favored over the reduction of the quark mass inside the soliton. However, the Gell-Mann Lévy model can only bind the quarks. It does not confine them. In some confining models, such as chirally symmetric color dielectric models [180], it is the reduction of the quark mass (bag formation) which is energetically favored. Solitons calculated in the Nambu Jona-Lasinio model have been unable to address the problem so far. Indeed, they have been performed using proper-time regularization, in which case the chiral field is forced to remain on the chiral circle in order to avoid an instability against collapse [153, 154] (see section 12.9).

12.1 Spectrum of u and d quarks interacting with a hedgehog field

A time-independent chiral field with a *hedgehog shape* can be written in the form:

$$S(\boldsymbol{r}) + i\gamma_5 P_a(\boldsymbol{r})\tau_a = M(r)\, e^{i\gamma_5 \hat{r}_a \tau_a \theta(r)} = M(r)\left[\cos\theta(r) + i\gamma_5 \widehat{r}_a \tau_a \sin\theta(r)\right] \tag{12.1}$$

In this expression, τ_a are the *three* Pauli matrices displayed in (A.21) and $\widehat{r}$ is the unit vector $\boldsymbol{r}/r$. The hedgehog configuration (12.1) is of particular interest because it constitutes a stationary point of the Nambu Jona-Lasinio action (4.23), the Gell-Mann Lévy action (11.10) and the Skyrmion action (12.9). Since $\widehat{r}_a$ and τ_a are vectors in respectively configuration and isospin space, the hedgehog field (12.1) is invariant in a simultaneous rotation of configuration space and isospin space. In section 13.3 we show that this symmetry leads to specific rotational bands with equal spin J and isospin T. The nucleon $N(939)$ with $J = T = 1/2$ and the delta $\Delta(1232)$ with $J = T = 3/2$ are members of such a rotational band.

The hedgehog field, which is used to describe baryons, is subject to the boundary conditions:

$$\theta(r=0) = n\pi \qquad \theta(r \to \infty) = 0 \qquad M(r \to \infty) = M \tag{12.2}$$

where n is an integer, called the *winding number* of the chiral field. The function $\theta(r)$ is often referred to as the *profile function* of the hedgehog field. As $r \to \infty$, the function $M(r) \to M$, where the constant M is the constituent quark mass. In the vacuum, at $r \to \infty$, the chiral field becomes $S = M$ and $P = 0$ and the spectrum displays a positive energy as well as a negative energy continuum beginning at $\pm M$, characteristic of a free Dirac particle of mass M.

The spectrum of quarks interacting with a time-independent hedgehog field can be calculated by searching for the eigenstates $|\lambda\rangle$ of the Dirac hamiltonian h:

$$h = \frac{\boldsymbol{\alpha}.\boldsymbol{\nabla}}{i} + \beta M(r)\left[\cos\theta(r) + i\gamma_5 \widehat{r}_a \tau_a \sin\theta(r)\right] \qquad h|\lambda\rangle = e_\lambda |\lambda\rangle \tag{12.3}$$

The orbits $|\lambda\rangle$ bear quantum numbers which reflect the symmetries of the Dirac hamiltonian (12.3). Let $\boldsymbol{j}$ and $\boldsymbol{t} = \boldsymbol{\tau}/2$ be the angular momentum and isospin operators acting on the single quark states $|\lambda\rangle$. The Dirac hamiltonian h commutes with the *grand spin* operator:

$$\boldsymbol{G} = \boldsymbol{j} + \boldsymbol{t} \tag{12.4}$$

which has commutation rules of an angular momentum. Eigenvectors of h can therefore be constructed by adding the angular momentum j to the isospin $t = 1/2$ so as to make a state with grand spin $G = j+1/2$ or $G = j-1/2$. The angular momentum $j = l \pm 1/2$ is obtained by adding the orbital angular momentum l to the spin $s = 1/2$ of the quark. The Dirac hamiltonian (12.3) commutes with the parity operator. For a given angular momentum j, we therefore have two orbits, one with parity $(-)^l = (-)^{j+1/2}$ and another with opposite parity $(-)^l = (-)^{j-1/2}$. The most commonly used method of calculating the eigenstates of quarks in a hedgehog field consists in diagonalizing the hamiltonian (12.3) in a basis of spherical plane waves normalized inside a spherical box [189]. The same method has been extended to calculate quark orbits in a sphaeleron [190, 191]. Other methods, involving phase shifts and Green's functions, have also been devised [192, 193].

The spectrum of orbits in a hedgehog field with winding number $n = 1$ is shown in Fig.12.1. The function $M(r)$ is assumed to remain a constant, independent of r and the profile function is assumed to have an exponential shape $\theta(r) = \pi e^{-\frac{r}{R}}$ where the radius R is a measure of the size of the system. In this case, we have

$$S^2(r) + P_a^2(r) = M^2 \tag{12.5}$$

and we say that the chiral field is restricted to the chiral circle. It is simple to check that the energies e_λ are proportional to M and that the ratio e_λ/M depends only on the dimensionless parameter MR. In figure 12.1, the energies e_λ/M of the orbits are plotted as a function of the dimensionless parameter MR. When $MR \to 0$ the chiral field loses all its effect (as a potential with a vanishing volume integral) and the energies of the orbits reduce to their unperturbed vacuum values.

As R grows, the chiral field is progressively switched on. A 0^+ orbit, with grand spin $G = 0$ and positive parity, is seen to come down from the positive energy continuum and to enter the mass gap. The negative energy orbits belonging to the Dirac sea do not deviate significantly from the negative energy continuum. Orbits with energies within the mass gap $-M < e_\lambda < M$ are *bound* states of the Dirac hamiltonian. They decay exponentially as $r \to \infty$. The spin and isospin wavefunction of a 0^+ orbit is fixed. The Pauli principle therefore allows only N_c quarks in this orbit, in which case the state is antisymmetric with respect to permutations of the color labels. It is a color singlet. The state with N_c quarks filling the 0^+ orbit has baryon number $B = 1$ (relative to the vacuum).

Figure 12.2 shows the spectrum of the Dirac hamiltonian when the chiral field has a hedgehog shape with a winding number $n = 2$. The profile function is $\theta(r) = 2\pi e^{-\frac{r}{R}}$. Two bound 0^+ orbits are seen to enter the mass gap. When the two 0^+ orbits are filled with N_c quarks each, a colorless state with baryon number $B = 2$ is formed. This is an illustration of the Atiyah–Singer index theorem [194, 195, 196], which in this case states that the number of bound

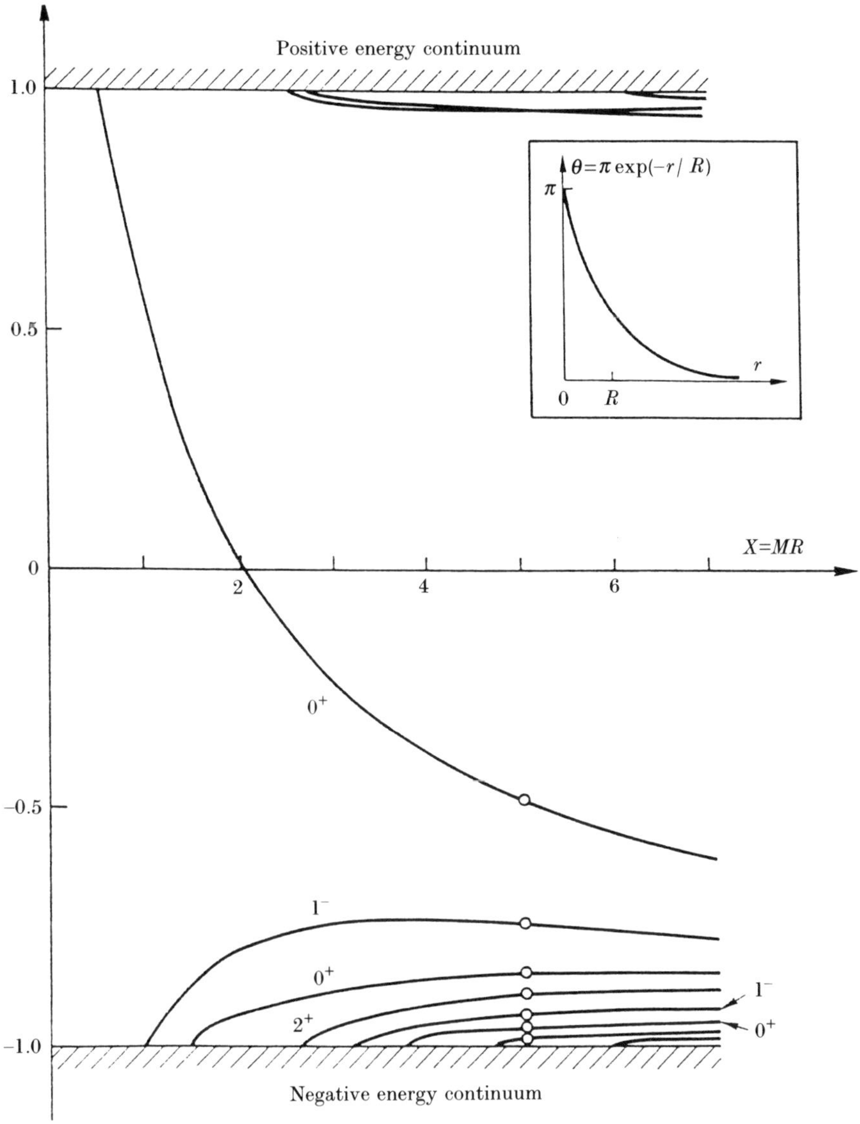

FIG. 12.1. Spectrum of the Dirac hamiltonian (12.3) plotted as a function of the dimensionless size parametre $X = MR$. The chiral field has winding number $n = 1$. The profile function $\theta(r)$ is shown in the inset. When the soliton has baryon number $B = 1$, the orbits marked with an open circle are filled with quarks, the others remaining empty.

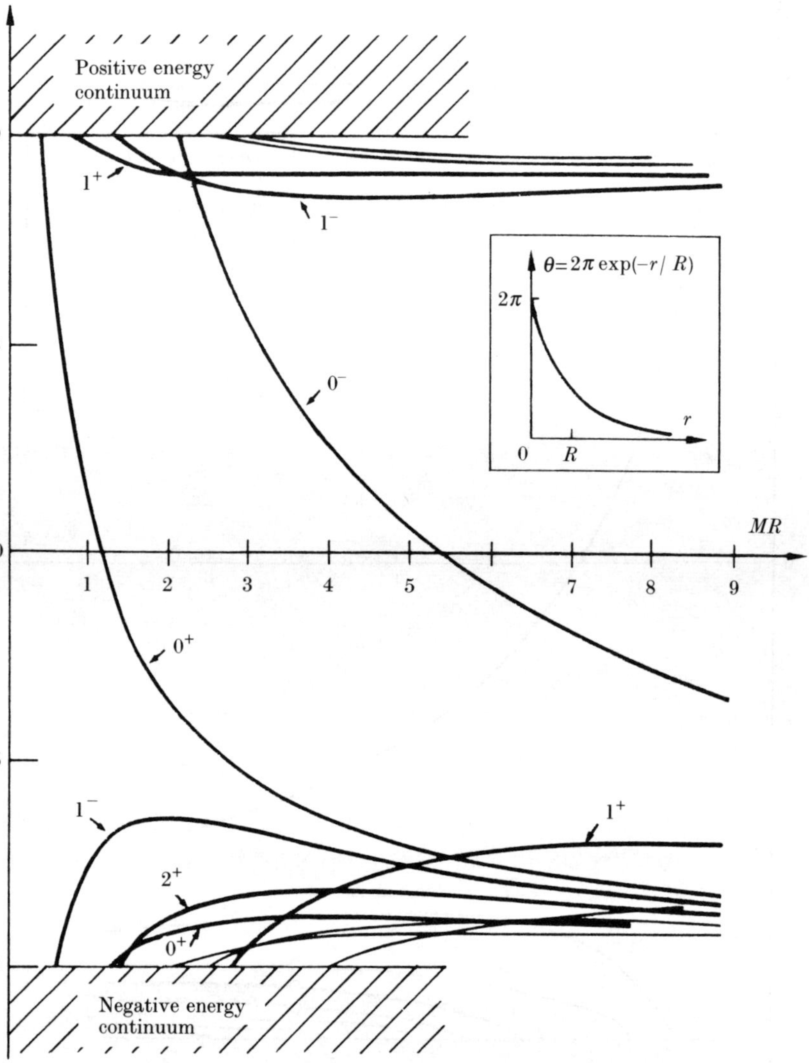

FIG. 12.2. The same as Fig. 12.1 for a chiral field with winding number $n = 2$. A soliton with baryon number $B = 2$ can be formed by filling the 0^+ and 0^- valence orbits, in addition to the remaining negative energy orbits.

states is equal to the winding number, so that the latter can be identified with the baryon number.

The winding number of the chiral field can thus distinguish systems with different baryon number. They are usually referred to as sectors. The *meson sector* has winding number $n = 0$ and the *baryon sector* has winding number $n = 1$. The winding number has a topological meaning [197, 198]. In Skyrme models, such as (12.9) for example, the different sectors are separated by infinite energy barriers and the solitons have topological stability. In chiral quark models, such as the Gell-Mann Lévy and Nambu Jona-Lasinio models, sectors with different winding numbers are separated by finite energy barriers. They give rise to so-called non-topological solitons.

12.2 Skyrme's topological density

Let $U(x) = e^{i\theta_a(x)\tau_a}$ be a chiral field defined on a chiral circle of unit radius. Skyrme suggested that the current:

$$j^\mu(x) = \frac{1}{24\pi^2}\varepsilon^{\mu\nu\alpha\beta} tr_f(L_\nu L_\alpha L_\beta) \qquad L_\mu = U^\dagger(\partial_\mu U) \qquad U(x) = e^{i\theta_a(x)\tau_a} \tag{12.6}$$

should be identified with the baryon current [197]. The current (12.6) is conserved: $\partial_\mu j^\mu = 0$. When the chiral field has a time-independent hedgehog shape $U(\boldsymbol{r}) = e^{i\hat{r}_a\tau_a\theta(r)}$, with a profile function which satisfies the boundary condition (12.2), it may be shown that the baryon number is equal to the winding number n of the chiral field:

$$B = \int d_3r\; j^{\mu=0}(\boldsymbol{r}) = n \tag{12.7}$$

This property is related to the topology of the mapping of $SU(2)$ or S^3 onto R^3, which is realized by the matrix $U(\boldsymbol{r})$. This is why the current (12.6) is often referred to as Skyrme's topological current.

We can compare the baryonic density $\rho(\boldsymbol{r}) = j^0(\boldsymbol{r})$, obtained from the topological current (12.6), with the quark density which is obtained from quarks interacting with the *same* chiral field. (Skyrme's topological current (12.6) can equally well be defined with a γ_5 in the definition of the chiral field.) The quark density may be computed from the quark orbits (12.3):

$$\rho(\boldsymbol{r}) = \sum_{\lambda_{occ}} \langle\lambda \mid \boldsymbol{r}\rangle\langle\boldsymbol{r} \mid \lambda\rangle - \sum_{e^0_\lambda<0} \langle\lambda_0 \mid \boldsymbol{r}\rangle\langle\boldsymbol{r} \mid \lambda_0\rangle \tag{12.8}$$

In this expression $\sum_{\lambda_{occ}}$ is a sum over all the occupied quark orbits, namely the negative energy Dirac sea orbits *and* the valence 0^+ orbit which lies within the mass gap, as shown in Fig.12.1. The second term of (12.8) is the unperturbed density of the Dirac sea, calculated with $S = M$ and $P = 0$. Figure 12.3, taken from [189], shows various contributions to the quark density, plotted as a function

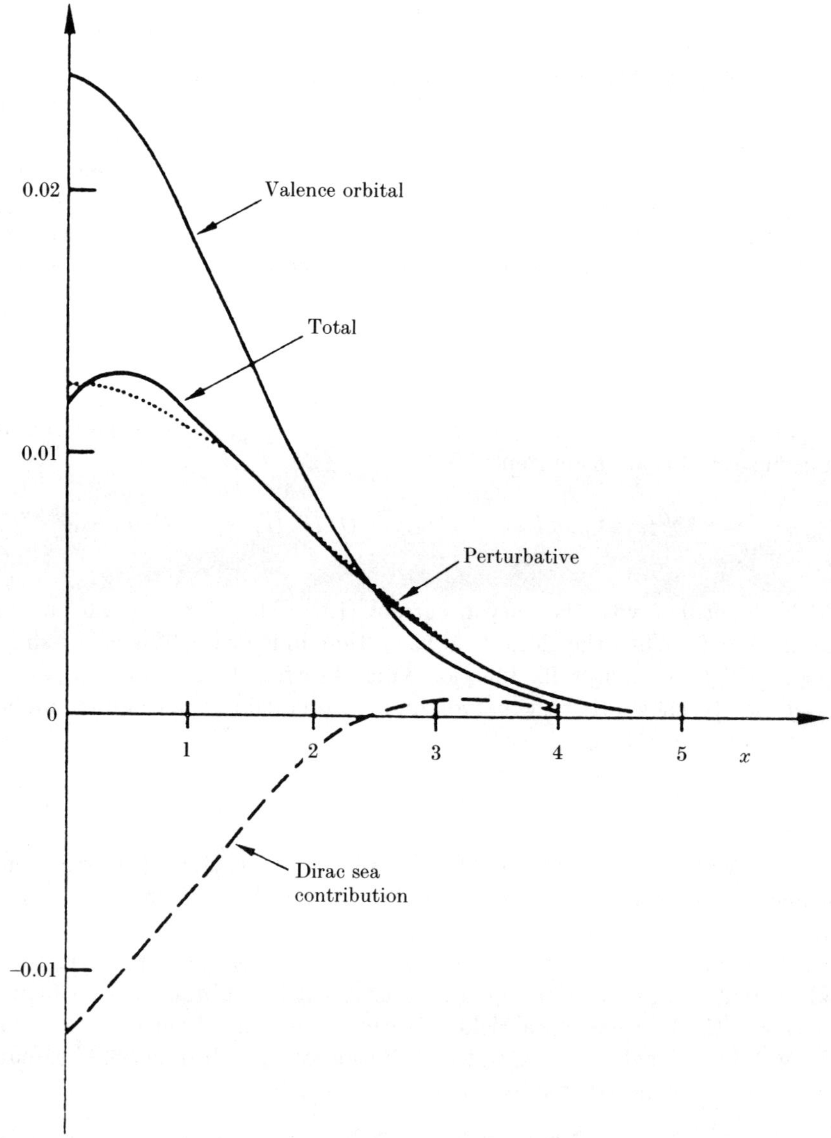

FIG. 12.3. The quark density (12.8) is plotted as a function of the dimensionless $x = Mr$ of the profile function. Valence and Dirac sea contributions are distinguished. The dashed curve is the density calculated from the Skyrme topological current (12.6). From [189].

of the dimensionless distance $x = Mr$. The density is calculated with a linear profile function $\theta(r) = \pi\left(1 - \frac{r}{R}\right)$ for $r < R$ and $\theta(r) = 0$ for $r > R$. The size parameter $MR = 5$ is typical of a soliton equilibrium size.

The quark density (12.8) (labeled 'Total' on the figure) is seen to be quite similar to the quark density obtained from Skyrme's topological current (12.6) (labeled 'Perturbative' on the figure). This is not entirely surprising, because the leading order term in the gradient expansion of the quark current $\langle\bar{q}\gamma^{\mu}q\rangle$ is exactly equal to Skyrme's topological current (12.6) [199, 200]. For smaller values of MR, the agreement between the densities deteriorates [189].

The comparison of quark densities is simply a property of the quark determinant $-Tr\ln(\partial_{\tau} + h)$, with h given by (12.3). It does not imply that the Skyrmion *dynamics* are the same as the dynamics of solitons described by the Nambu Jona-Lasinio or the Diakonov Petrov models for example. In the chiral limit, the Skyrme lagrangian is [197]:

$$\mathcal{L} = -\frac{f_{\pi}^{2}}{4}tr_{f}\left(L_{\mu}L^{\mu}\right) + \frac{1}{32e^{2}}tr_{f}\left[L_{\mu}, L_{\nu}\right]\left[L^{\mu}, L^{\nu}\right] \tag{12.9}$$

where e^2 is a dimensionless constant and where L_{μ} is defined in (12.6). Skyrmion dynamics are reviewed in [201, 202] . The Nambu Jona-Lasinio model has an extra degree of freedom because the chiral field $S + i\gamma_5 P_a\tau_a$ is allowed to stray off the chiral circle $S^2 + P_a^2 = M^2$. But even the Diakonov Petrov model (11.2), in which the chiral field is constrained to the chiral circle, differs from the Skyrmion. Indeed, the gradient expansion of the quark determinant $-Tr\ln(\partial_{\tau} + h)$ gives rise to several terms which are fourth order in the derivatives of U [200], only one of which appears in the Skyrme lagrangian (12.9). In fact, models based on gradient expansions of the quark determinant fail to yield stable solitons [124].

The Skyrme lagrangian (12.9) originated before the discovery of quarks and indeed, quarks need never be mentioned in Skyrmion physics. However, the fact that the Skyrme lagrangian needs to be supplemented by a Wess–Zumino term, the normalization of which coincides with the leading order contribution of the quark determinant [198, 200, 203], makes one strongly suspect that the underlying physics of the Skyrmion is a model of quarks interacting with a chiral field. Gradient expansions of the quark determinant do not exhibit the lack of confinement of chiral quark models. Therefore gradient expansions of the quark determinant necessarily break down at the energies at which free quarks can propagate. This corresponds to energies close to and above $2M$ in the meson $q\bar{q}$ sector and to $N_c M$ in the soliton sector. It does not make sense to use gradient expansions of quark determinants at such energies, so as to identify axial meson masses or to calculate loosely bound solitons for example. For these reasons, Skyrme-like models cannot be justified in terms of quarks interacting with chiral fields. They have quite different properties. For example, Skyrme-like models have infinite energy barriers separating sectors with different winding numbers and the solitons they give rise to are topologically stable. Chiral quark models, such as the Nambu Jona-Lasinio, Gell-Mann Lévy or Diakonov Petrov models have finite energy barriers separating the different sectors (although they strictly conserve baryon number) and they give rise to so-called non-topological solitons.

12.3 The soliton calculated with proper-time regularization

In this section we show how solitons are calculated in the Nambu Jona-Lasinio model using the proper-time regularization (section 6.3.1), which is the most commonly used regularization. We start with the expression (4.20) of the Nambu Jona-Lasinio euclidean action. We neglect vector interactions, which are discussed in section 12.10. A soliton is a localized time-independent stationary point of the action. It is a bound state of quarks. When the fields are time-independent, the Dirac hamiltonian h, defined in (4.17), commutes with the time derivative ∂_τ . We can easily check that we then have $-Tr\ln(\partial_\tau + h) = -Tr\ln(-\partial_\tau + h)$. It follows that, for time-independent fields, the imaginary part of the action vanishes. The proper-time regularized action (6.18) is thus:

$$I(S,P) = \frac{1}{2}Tr\int_{\frac{1}{\Lambda^2}}^{\infty}\frac{ds}{s}e^{-s\left(-\partial_\tau^2+h^2\right)} + \frac{1}{2g_s^2}\int d_4x\left(S_a^2+P_a^2\right) \tag{12.10}$$

with:

$$h = \frac{\boldsymbol{\alpha}.\boldsymbol{\nabla}}{i} + \beta\left(m + S_a\tau_a + i\gamma_5 P_a\tau_a\right) \tag{12.11}$$

The trace in the regularized action (12.10) can be evaluated in the basis $|\omega,\lambda\rangle$ of the common eigenstates of ∂_τ and h:

$$\partial_\tau|\omega,\lambda\rangle = i\omega|\omega,\lambda\rangle \qquad h|\omega,\lambda\rangle = e_\lambda|\omega,\lambda\rangle \qquad \langle\tau\mid\omega,\lambda\rangle = \frac{1}{\sqrt{\beta}}e^{i\tau\omega}|\lambda\rangle \tag{12.12}$$

The action (12.10) becomes:

$$I(S,P) = \frac{1}{2}\sum_{\omega\lambda}\int_{\frac{1}{\Lambda^2}}^{\infty}\frac{ds}{s}e^{-s\left(\omega^2+e_\lambda^2\right)} + \frac{\beta}{2g_s^2}\int d_3r\left(S_a^2+P_a^2\right)$$

$$= \frac{1}{2}\frac{\beta}{2\pi}\sqrt{\pi}\sum_{\lambda}\int_{\frac{1}{\Lambda^2}}^{\infty}\frac{ds}{s^{3/2}}e^{-se_\lambda^2} + \frac{\beta}{2g_s^2}\int d_3r\left(S_a^2+P_a^2\right) \tag{12.13}$$

where we performed the sum over ω in the zero temperature limit: $\sum_\omega \to \frac{\beta}{2\pi}\int_{-\infty}^{\infty}d\omega$. No confusion should arise from the use of β to denote the inverse temperature in equations (12.12) and (12.13) and the Dirac matrix $\beta = \gamma^0$ in equations (12.11) and (12.16) for example. The energy of the system is $E(S,P) = -(1/\beta)\, I(S,P)$. Integrating s by parts, we find:

$$E(S,P) = \sum_{\lambda}\left(\frac{1}{2\sqrt{\pi}}\Lambda e^{-\frac{e_\lambda^2}{\Lambda^2}} - \frac{1}{2}e_\lambda n_\lambda\right) + \sum_{e_\lambda<0} e_\lambda\, n_\lambda + \frac{1}{2g_s^2}\int d_3r\left(S_a^2 + P_a^2\right) \tag{12.14}$$

where n_λ is a normalization, assigned to the quark orbit $|\lambda\rangle$ by the proper-time regularization:

$$n_\lambda = \frac{|e_\lambda|}{\sqrt{\pi}}\int_{\frac{1}{\Lambda^2}}^{\infty}\frac{e^{-se_\lambda^2}}{\sqrt{s}}ds = \frac{1}{\sqrt{\pi}}\int_{\frac{e_\lambda^2}{\Lambda^2}}^{\infty}\frac{e^{-x}}{\sqrt{x}}dx \underset{\Lambda\to\infty}{\to} 1 \tag{12.15}$$

In section 12.5 we show that, in some cases, a valence contribution (12.34) needs to be added to the energy $E(S,P)$. In the limit $\Lambda \to \infty$ of an infinite cut-off, the incongruous-looking first term of (12.14) becomes an infinite but irrelevant constant independent of the fields (S,P), because $\sum_\lambda e_\lambda = Trh = 0$.

The field configuration which minimizes the energy with respect to the fields S and P is also a stationary point of the action. It is the solution of the equations $\delta E/\delta S_a(\boldsymbol{r}) = 0$ and $\delta E/\delta P_a(\boldsymbol{r}) = 0$, which can be made more explicit using (12.14):

$$-\frac{1}{2}\sum_{\lambda} n_\lambda \frac{e_\lambda}{|e_\lambda|}\langle\lambda\mid\boldsymbol{r}\rangle\,\beta\tau_a\,\langle\boldsymbol{r}\mid\lambda\rangle + \frac{1}{g_s^2}S_a(\boldsymbol{r}) = 0$$

$$-\frac{1}{2}\sum_{\lambda} n_\lambda \frac{e_\lambda}{|e_\lambda|}\langle\lambda\mid\boldsymbol{r}\rangle\, i\beta\gamma_5\tau_a\,\langle\boldsymbol{r}\mid\lambda\rangle + \frac{1}{g_s^2}P_a(\boldsymbol{r}) = 0 \tag{12.16}$$

In section 12.5 a valence contribution is added to the left hand sides of these equations. For a soliton, which is a localized solution, the fields must obey the boundary conditions:

$$S_a(\boldsymbol{r}) \underset{r\to\infty}{\to} M_a \qquad\qquad P_a(\boldsymbol{r}) \underset{r\to\infty}{\to} 0 \tag{12.17}$$

which state that, far from the soliton, the fields acquire their vacuum values. The latter are determined by the gap equations (5.14).

The equations (12.11), (12.3) and (12.16) constitute a set of coupled non-linear equations with which the eigenstates $|\lambda\rangle$ and the eigenvalues e_λ of the Dirac hamiltonian (12.11) can be calculated. They can be solved by iteration. An initial guess is made for the chiral field. It could be, for example, the hedgehog shape (12.3) with a winding number $n = 1$, with a linear or gaussian decreasing profile function $\theta(r)$ and with a constant constituent quark mass M. The initial guess defines a first approximation for the Dirac hamiltonian (12.11). Its eigenvectors and eigenvalues can then be calculated as in equation (12.3). The most commonly used method consists in diagonalizing it in a basis of spherical Dirac plane waves [189]. The eigenvalues e_λ and the eigenvectors $|\lambda\rangle$ can then be used to calculate the chiral field, according to equations (12.16). The process is then iterated: the calculated chiral field defines a next approximation for the Dirac hamiltonian (12.11), which is used to recalculate the eigenvectors according to (12.3), and so on until successive iterations yield similar eigenvectors and eigenvalues of h.

An alternative way to proceed is to minimize the energy (12.14) with respect to some assumed shapes of the fields $S_a(\boldsymbol{r})$ and $P_a(\boldsymbol{r})$. This is particularly useful when approximate constrained variations are considered. For example, the chiral angle may be chosen to have the exponential form $\theta(r) = \pi e^{-r/R}$ and the energy is then minimized with respect to R. Or, the function $M(r)$ may be kept constant, so as to constrain the fields to the chiral circle. However, such procedures are blind to the existence of self-consistent symmetries (section 12.6) and they may not lead to a true stationary point of the action.

The energy of the soliton is the difference between the energy calculated with the fields obtained by the iteration process described above, and the vacuum energy:

$$E_{soliton} = E(S,P) - E(S = M, P = 0) \tag{12.18}$$

where $E(S,P)$ is the functional (12.14) and where $S_a = M_a$, $P_a = 0$ are the vacuum asymptotic values (12.17). A necessary (but not sufficient) condition for the iteration process to converge is that the soliton should be a *bound state* of quarks. This means that the energy (12.18) of the soliton should be lower than the sum of its constituent quark masses. Otherwise the solution would either fly apart when successive iterations are performed, or it would represent an unstable solution, because energy could be gained by separating the soliton into its constituent quarks. The energy of the soliton is lowered when the chiral field (with a hedgehog shape) forms a bound quark orbit, such as the 0^+ orbit displayed in Fig. 12.1. However, some of this binding energy is lost because the chiral field distorts the Dirac sea. In practice, the two processes almost compensate each other and the binding energy of the soliton, which is the difference between the soliton energy (12.18) and the masses M of its constituents, is quite small, rarely exceeding 15% of the soliton energy. An example is given in Table 12.1. Further problems, related to the stability of the soliton, are discussed in section 12.9.

12.4 Hermitian and antihermitian probes

The proper-time regularization regularizes the real part of the action but not the imaginary part. As a result, perturbations are not analytic in the coupling constants. To illustrate this and to collect results which will be useful in later applications, we consider adding to the Dirac hamiltonian a time-independent probe $f+ig$, where f and g are hermitian operators acting on the quark variables. In the presence of the probes, we have:

$$D^\dagger D = (\partial_\tau + h + f - ig)(-\partial_\tau + h + f + ig) = D_0^\dagger D_0 + V \tag{12.19}$$

where:

$$D_0^\dagger D_0 = -\partial_\tau^2 + h^2$$

$$V = \{f, h\}_+ + i\{\partial_\tau, g\} + i[h, g] + (f - ig)(f + ig) \tag{12.20}$$

Let us expand the proper-time regularized action to second order in the probe. We use the Dyson expansion:

$$e^{-s\left(D_0^\dagger D_0 + V\right)}$$

$$= e^{-sD_0^\dagger D_0}\left(1 - \int_0^s du V(u) + \int_0^s du \int_0^u dv V(u) V(v) - ...\right) \tag{12.21}$$

where:

$$V(u) = e^{uD_0^\dagger D_0} V e^{-uD_0^\dagger D_0} \tag{12.22}$$

A straightforward calculation will then yield the following expansion of the real part of the proper-time regulated quark determinant:

$$-\frac{1}{2}Tr_\Lambda \ln\left(D_0^\dagger D_0 + V\right) = \frac{1}{2}Tr\int_{\frac{1}{\Lambda^2}}^\infty \frac{ds}{s} e^{-s\left(D_0^\dagger D_0 + V\right)} = \frac{1}{2}Tr\int_{\frac{1}{\Lambda^2}}^\infty \frac{ds}{s} e^{-s\left(D_0^\dagger D_0\right)}$$

$$-\frac{1}{2}Tr\int_{\frac{1}{\Lambda^2}}^\infty ds\, e^{-sD_0^\dagger D_0} V + \frac{1}{4}Tr\int_{\frac{1}{\Lambda^2}}^\infty sds \int_0^1 dz e^{-sD_0^\dagger D_0(1-z)} V e^{-sD_0^\dagger D_0 z} V + ... \tag{12.23}$$

This expansion, together with the basis states (12.12), allows us to expand $-\frac{1}{2}Tr_\Lambda \ln\left(D^\dagger D\right)$ in powers of f and g, in the zero temperature limit. The result can be written:

$$-\frac{1}{2}Tr_\Lambda \ln\left(D^\dagger D\right) = \frac{\beta}{4\sqrt{\pi}}\sum_\lambda \int_{\frac{1}{\Lambda^2}}^\infty \frac{ds}{s^{3/2}}\, e^{-se_\lambda^2} + R^f + R^g + R^{ff} + R^{gg} + R^{fg} \tag{12.24}$$

where $R^f, R^{ff}, ...$ are the contributions which are first order in f, second order in f, and so on. The first order contribution to the real part comes from the hermitian probe f only:

$$R^f = -\frac{\beta}{2}\sum_\lambda n_\lambda \frac{e_\lambda}{|e_\lambda|}\langle\lambda|f|\lambda\rangle \underset{\Lambda\to\infty}{\to} \beta\sum_{e_h<0}\langle h|f|h\rangle - \frac{\beta}{2}trf \qquad R^g = 0 \tag{12.25}$$

In (12.25), n_λ is the normalization defined in (12.15). The equations (12.16) are special cases of (12.25). The second order contributions to the real part of the action are:

$$R^{ff} = -\frac{\beta}{4}\sum_{\lambda\mu}\frac{\langle\lambda|f|\mu\rangle\langle\mu|f|\lambda\rangle}{e_\lambda - e_\mu}\left(n_\lambda\frac{e_\lambda}{|e_\lambda|} - n_\mu\frac{e_\mu}{|e_\mu|}\right)\underset{\Lambda\to\infty}{\to} -\beta\sum_{\substack{e_p>0\\e_h<0}}\frac{\langle p|f|h\rangle\langle h|f|p\rangle}{e_p - e_h}$$

$$R^{gg} = -\frac{\beta}{4}\sum_{\lambda\mu}\frac{\langle\lambda|g|\mu\rangle\langle\mu|g|\lambda\rangle}{e_\lambda - e_\mu}\left(2\Lambda\frac{e^{-\frac{e_\lambda^2}{\Lambda^2}} - e^{-\frac{e_\mu^2}{\Lambda^2}}}{e_\lambda^2 - e_\mu^2} - \left(n_\lambda\frac{e_\lambda}{|e_\lambda|} - n_\mu\frac{e_\mu}{|e_\mu|}\right)\right)$$

$$\underset{\Lambda\to\infty}{\to}\beta\sum_{\substack{e_p>0\\e_h<0}}\frac{\langle p|g|h\rangle\langle h|g|p\rangle}{e_p - e_h}$$

$$R^{fg} = 0 \tag{12.26}$$

In the limit $\Lambda \to \infty$, we have $R^{ff} = -R^{gg}$. In the expressions (12.25) and (12.26), we adopted the *particle–hole* notation:

$$|\lambda\rangle = |p\rangle \quad \text{if} \quad e_\lambda > 0 \qquad\qquad |\lambda\rangle = |h\rangle \quad \text{if} \quad e_\lambda < 0 \tag{12.27}$$

In section 12.5 valence corrections are added to the quantities (12.25) and (12.26). In the unregularized theory, the hole states $|h\rangle$ are filled in the zero temperature limit and the particle states $|p\rangle$ are empty. In (12.25), the trace is over a complete set of quark states:

$$trf = \sum_\lambda\langle\lambda|f|\lambda\rangle = \sum_{e_h<0}\langle h|f|h\rangle + \sum_{e_p>0}\langle p|f|p\rangle \tag{12.28}$$

We can also expand the imaginary part of the action in powers of f and g :

$$-\frac{1}{2}Tr\ln D + \frac{1}{2}Tr\ln D^\dagger = -iTr\frac{1}{\partial_\tau + h}g + iTr\frac{1}{\partial_\tau + h}g\frac{1}{\partial_\tau + h}f = I^g + I^{fg} \tag{12.29}$$

The terms I^f, I^{ff} and I^{gg} all vanish. The non-vanishing terms are:

$$I^g = i\beta\sum_{e_h<0}\langle h|g|h\rangle \qquad\qquad I^{fg} = -i\beta\sum_{\substack{e_p>0\\e_h<0}}\frac{\langle p|g|h\rangle\langle h|f|p\rangle + \langle p|f|h\rangle\langle h|g|p\rangle}{e_p - e_h} \tag{12.30}$$

The contributions to the imaginary part of the action are not regularized. Valence corrections to these quantities are given in section (12.5). The first order results above show that, if we probe the system with a hermitian operator f, then the expectation value $\langle f\rangle$ is the regularized expression R^f given by (12.25). If we probe it with an antihermitian operator if, we find the unregularized expression (12.30), so that we do *not* find that $\langle if\rangle = i\langle f\rangle$. Instead, $\langle if\rangle = i\sum_{e_h<0}\langle h|f|h\rangle$, which can be divergent. The proper-time regularized quark determinant is thus a non-analytic, discontinuous and singular functional of the probes.

12.5 The baryon number and valence orbits

What is the baryon number of the system which is described by the action (12.13) or the energy (12.14)? Consider adding a hermitian probe to the system, which has the form of a quark bilinear operator $\varepsilon F = \varepsilon q^{\dagger} f q$, where ε is an infinitesimal number and where f is a hermitian operator acting on the quark variables. The effect is to modify the Dirac hamiltonian (12.11), which becomes:

$$h \to h + \varepsilon f \tag{12.31}$$

The expectation value of the operator F can be deduced from the expression (12.25):

$$\langle F \rangle = \frac{1}{\beta}\frac{\delta}{\delta\varepsilon}\left(-Tr \ln\left(\partial_\tau + h + \varepsilon f\right)\right) = -\frac{1}{2}\sum_\lambda n_\lambda \frac{e_\lambda}{|e_\lambda|}\langle \lambda \,|f|\, \lambda\rangle \tag{12.32}$$

$$= \sum_{e_h<0} n_h \langle h\,|f|\,h\rangle - \frac{1}{2}\sum_\lambda n_\lambda \langle \lambda\,|f|\,\lambda\rangle \underset{\Lambda\to\infty}{\to} \sum_{e_h<0} \langle h\,|f|\,h\rangle - \frac{1}{2} trf \tag{12.33}$$

Consider first the case $\Lambda \to \infty$ of an infinite cut-off. The term $-\frac{1}{2}trf$ is a constant which vanishes when the difference between expectation values $\langle F \rangle$ is taken between two states, such as the soliton and the vacuum. The expression (12.33) shows that the occupied orbits $|\lambda\rangle$ are the negative energy orbits with $e_\lambda < 0$. This is to be expected because, in the zero temperature limit, the operator $e^{-\beta H}$ projects onto the ground state *in Fock space*, which, for a system of fermions (the quarks) interacting with an external field (the chiral field), is the state obtained by filling all the negative energy orbits. The number of quarks, which is equal to N_c times the baryon number, is obtained by setting $f = 1$ in the expression (12.33). It is equal to the number of negative energy orbits. In the vacuum, $S = M$ and $P = 0$. The spectrum of h is composed of plane wave states with equal and opposite energies. The ground state is well defined: it consists of a Dirac sea of quarks filling the negative energy plane wave states. However, when the chiral field acquires a hedgehog shape (12.1) with winding number $n = 1$ for example, one 0^+ orbit enters the mass gap as shown in Fig. 12.1. We shall refer to this orbit as the *valence orbit*. With higher winding numbers, several valence orbits would occur. Depending on the size of the soliton, the orbit can have positive or negative energy. When the orbit crosses the zero energy line, nothing much happens to the system, except that the ground state of the system, as viewed in Fock space, now includes quarks occupying the 0^+ orbit. The baryon number increases by one unit when a 0^+ orbit acquires a negative energy. This is more a question of bookkeeping than of dynamics. We want to calculate a soliton with baryon number $B = 1$, even when the valence 0^+ orbit has positive energy. We are less interested in calculating the ground state of the system in Fock space. For example, in Fock space it is the vacuum and not the nucleon $N\,(939)$ which is the ground state of a strongly interacting system.

The usual way of controlling the baryon number of the system consists in subtracting a chemical potential μ from the Dirac hamiltonian, which becomes $h-\mu$. When the chemical potential is chosen to be higher than the energy of the 0^+ orbit, the ground state of the system of quarks, calculated with the hamiltonian $h-\mu$, will contain the 0^+ orbit. With such a chemical potential, the negative energy hole states $|h\rangle$ include the valence orbit and the expression (12.33) will yield a baryon number $B=1$, after subtracting the baryon number of the vacuum, calculated with the vacuum values $S=M$, $P=0$ and $\mu=0$. However, problems arise when a chemical potential is used in proper-time regularization with a finite cut-off. The term $-\frac{1}{2}\sum_\lambda n_\lambda \langle\lambda\,|f|\,\lambda\rangle$ of (12.33) is then no longer equal to $tr\, f$ and its value is a continuous functional of the chiral field. If we set $f=1$ we find that the number of quarks is a non-integer continuous number. Furthermore, if we make the substitution $h\to h-\mu$ in the expression (12.14) of the energy, the first term includes a finite contribution and the shift in energy is not equal to $-\mu B$, as it would in the infinite cut-off limit. These are artifacts of the proper-time regularization, which favors one particular choice of the zero energy. For these reasons, a chemical potential is not used in conjunction with proper-time regularization. There seems to be no other way to calculate solitons with baryon number $B=1$ than to keep an eye on the energy of the 0^+ valence orbit. Whenever it has positive energy, it is filled with N_c quarks by hand, so to speak. One simply adds the contribution of the positive energy valence orbit.

This is done by first considering the infinite cut-off limit $\Lambda\to\infty$. Expectation values are then expressed in terms of the particle and hole orbits (12.27). The valence orbit (or orbits) $|val\rangle$ are recognized from the calculated spectrum of the Dirac hamiltonian h. *Positive energy valence orbits are then added, to the set of hole orbits.* Contributions from negative energy valence orbits need not be added, because they are already included in the expectation values derived in section 12.4. Since there is only a finite number of valence orbits, the valence corrections are finite. Thus, we add a valence contribution to the energy (12.14):

$$E_{val}(S,P)=\sum_{\{val\}}\theta(e_{val})\,e_{val} \qquad \theta(e_{val})=\begin{cases}1 & if \quad e_{val}>0\\ 0 & if \quad e_{val}<0\end{cases} \tag{12.34}$$

The equations (12.16) have corresponding valence corrections. For example, the first equation (12.16) will read:

$$-\frac{1}{2}\sum_\lambda n_\lambda\frac{e_\lambda}{|e_\lambda|}\langle\lambda\,|\,\boldsymbol{r}\rangle\,\beta\tau_a\,\langle\boldsymbol{r}|\,\lambda\rangle+\sum_{\{val\}}\theta(e_{val})\,\langle val\,|\,\boldsymbol{r}\rangle\,\beta\tau_a\,\langle\boldsymbol{r}|\,val\rangle+\frac{1}{g_s^2}S_a(\boldsymbol{r})=0 \tag{12.35}$$

The expectation values of probes, listed in section 12.4, also require valence corrections. For example:

$$\delta I^{fg} = -i\beta\theta\left(e_{val}\right) \sum_{\substack{e_p>0 \\ \{val\}}} \frac{\langle p\,|g|\,val\rangle \langle val\,|f|\,p\rangle + \langle p\,|f|\,val\rangle \langle val\,|g|\,p\rangle}{e_p - e_h}$$

$$+i\beta\theta\left(e_{val}\right) \sum_{\substack{e_h<0 \\ \{val\}}} \frac{\langle val\,|g|\,h\rangle \langle h\,|f|\,val\rangle + \langle val\,|f|\,h\rangle \langle h\,|g|\,val\rangle}{e_p - e_h} \tag{12.36}$$

The second term of (12.36) appears because we wrote down the *correction* to I^{fg}. Similar valence corrections can be derived for all other quantities. It is not clear whether the normalization factors (12.15) should multiply the sums over particle and hole states appearing in the valence corrections of R^{ff} and R^{gg}.

Note that if we are only interested in evaluating the baryon density or the baryon number, we could probe the system with the antihermitian operator $f = i\,|\boldsymbol{r}\rangle\langle\boldsymbol{r}|$. Indeed, the antihermitian probe yields a density of the form I^g which is the same as that obtained with an infinite cut-off. When the valence 0^+ orbit has positive energy, its contribution to the baryon density would need to be added by hand.

12.6 Self-consistent symmetries of $SU(3)_f$ solitons

In this section we discuss self-consistent symmetries of solitons which are bound states of u, d and s quarks. Consider for example parity which transforms a quark orbit $\langle \boldsymbol{r} \mid \lambda\rangle$ into $\beta\langle -\boldsymbol{r} \mid \lambda\rangle$. The Dirac hamiltonian (12.11) commutes with parity, provided that the fields satisfy the conditions:

$$S_a(\boldsymbol{r}) = S_a(-\boldsymbol{r}) \qquad\qquad P_a(\boldsymbol{r}) = -P_a(-\boldsymbol{r}) \tag{12.37}$$

Parity is a self-consistent symmetry in the following sense. If we start the iteration process with fields which satisfy (12.37), then the Dirac hamiltonian (12.11) commutes with parity. The eigenstates $|\lambda\rangle$ of h can then be chosen with definite parity: $\langle \boldsymbol{r} \mid \lambda\rangle = \pm\beta\langle -\boldsymbol{r} \mid \lambda\rangle$. At the next iteration, the equations (12.16) will yield fields $S_a(\boldsymbol{r})$ and $P_a(\boldsymbol{r})$ which again satisfy the conditions (12.37). Orbits with definite parity are thus recovered at each iteration. Consider next symmetry with respect to rotations generated by the grand spin operator (12.4):

$$\boldsymbol{G} = \boldsymbol{j} + \frac{1}{2}\boldsymbol{\tau} \qquad\qquad \boldsymbol{\tau} = (\tau_1, \tau_2, \tau_3) \tag{12.38}$$

Assume that the u and d quarks have equal masses. The form (5.50) of the current quark mass term is then:

$$m = \frac{m_u + m_d + m_s}{\sqrt{6}}\tau_0 + \frac{m_u + m_d - 2m_s}{2\sqrt{3}}\tau_8 \tag{12.39}$$

It follows that m commutes with the three generators τ_1, τ_2 and τ_3 as well as with τ_8. Suppose that we start the iteration process with a chiral field which has the form of a trivial embedding of a $SU(2)$ hedgehog in $SU(3)$:

$$S_{a=0}(\boldsymbol{r}) = M_0(r) \qquad S_{a=8}(\boldsymbol{r}) = M_8(r) \qquad S_{a=1,\dots,7}(\boldsymbol{r}) = 0$$

$$P_{a=1,2,3}(\boldsymbol{r}) = \widehat{r}_a P(r) \qquad P_{a=0}(\boldsymbol{r}) = 0 \qquad P_{a=4,\dots,8}(\boldsymbol{r}) = 0 \tag{12.40}$$

The expression (13.42) is an example of such an embedding. The Dirac operator will then commute with both parity and the grand spin operator (12.38). Had we chosen $S_{a=1,2,3}(\boldsymbol{r}) = \widehat{r}_a S(r)$ or, for example, $P_{a=0}(\boldsymbol{r}) = P_0(r)$, the Dirac hamiltonian would not commute with the parity operator. When the fields have the form (12.40), the Dirac hamiltonian separates into two commuting terms $h_{(u,d)}$ and h_s. The term $h_{(u,d)}$ is a 2×2 matrix in flavor space, acting on the u and d quarks. The term h_s acts only on the s quark:

$$h = h_{(u,d)} + h_s \qquad [h_{u,d}, h_s] = 0$$

$$h_{(u,d)} = \frac{\boldsymbol{\alpha}.\boldsymbol{\nabla}}{i} + \beta\left(m + M(r) + i\gamma_5 \widehat{r}_a \tau_a P(r)\right) \qquad h_s = \frac{\boldsymbol{\alpha}.\boldsymbol{\nabla}}{i} + \beta\left(m_s + M_s(r)\right) \tag{12.41}$$

In (12.41), we use the notation (5.51) for the matrices m and M, which are diagonal in flavor space:

$$m + M = \sum_{i=u,d} |i\rangle (m+M) \langle i| \qquad (m_s + M_s) = |s\rangle (m_s + M_s) \langle s| \tag{12.42}$$

The eigenstates $\langle i = u,d \mid \lambda\rangle$ of the Dirac hamiltonian $h_{(u,d)}$ can be labeled $|n,\pi,j,G,M\rangle$, where π is the parity of the orbit, j the total angular momentum, $G = j \pm 1/2$ the total grand spin and M the projection of the grand spin. The additional (radial) quantum number $n = 1,2,3,\dots$ serves to distinguish orbits with the same quantum numbers π, j, G and M. Since h commutes with $\boldsymbol{G}$, orbits which differ only by the projection M of the grand spin are degenerate. It follows from equation (12.16) that $S_{a=0}$, $S_{a=8}$ and $\sum_{a=1}^{3} P_a \tau_a$ are invariant under rotations generated by the grand spin operator (12.38) because the orbits $|\lambda\rangle$ form *closed shells of grand spin.* This means that the proper-time normalization (12.15) of an orbit is independent of the projection M. The fields $S_{a=0}$, $S_{a=8}$ are therefore functions of $r = |\boldsymbol{r}|$ and $P_{a=1,2,3}$ transforms as the vector $\widehat{r}_a$, as indicated in (12.40). Invariance under rotations generated by the grand spin $\boldsymbol{G}$ follows from the invariance of the projector $\sum_M |GM\rangle \langle GM|$. Similarly, $P_{a=0}(\boldsymbol{r})$ and $P_{a=8}(\boldsymbol{r})$ have to be scalars of the form $P_{a=0}(r)$ and $P_{a=8}(r)$. However, they must vanish because they have to have odd parity. The same reasoning shows that the eigenstates of h_s can be labelled $|s,n,\pi,j,G,M\rangle$. Since the Dirac hamiltonian does not have matrix elements connecting the fields with flavor indices $i = 1,2$ and $j = 3$, once the fields $S_{a=4,\dots,7}(\boldsymbol{r})$ and $P_{a=4,\dots,7}(\boldsymbol{r})$ are set to zero, they remain zero at each successive iteration. Thus, at each iteration, the equations (12.16) will yield fields of the hedgehog form (12.40). This shows that invariance with respect to grand spin rotations is a self-consistent symmetry and that a stationary point of the energy exists, which has the hedgehog shape (12.40). Self-consistent symmetries are also encountered in Hartree–Fock theory [42]. The arguments above do not prove that solutions with other self-consistent symmetries cannot be constructed. For example, we could try a solution of the form $S_{ij}(\boldsymbol{r}) = \delta_{ij} a(r)\cos\alpha + \widehat{r}_i \widehat{r}_j b(r)\left(e^{i\gamma} - \cos\alpha\right) + \varepsilon_{ijk}\widehat{r}_k c(r)\sin\alpha$ [204]. So far, no solution with baryon number $B = 1$ has been reported to be energetically favored over the hedgehog solution (12.40). Solutions with different symmetries could however be energetically favored for excited states with various baryon numbers.

12.7 Spontaneously broken symmetries

Some symmetries of the action, such as translational symmetry, are always broken in localized soliton solutions. They give rise to three spurious states (section 8.4). Although symmetry with respect to rotations generated by the grand spin operator $\boldsymbol{G} = \boldsymbol{j} + \frac{1}{2}\boldsymbol{\tau}$ is not broken, symmetry with respect to rotations generated by the separate operators $\boldsymbol{j}$ and $\frac{1}{2}\boldsymbol{\tau}$ are broken. Because they are not independent, they give rise to three spurious states, as well as to collective rotations discussed in chapter 13.

12.8 $SU(2)_f$ solitons in the proper-time regularized Diakonov Petrov model

We illustrate solitons obtained with an $SU(2)_f$ hedgehog shaped chiral field by considering the Diakonov Petrov model [152], described by the action (11.2). In this model, the chiral field is restricted to the chiral circle. A hedgehog solution may be found in which the Dirac hamiltonian has the form (12.41) with:

$$h_{(u,d)} = \frac{\boldsymbol{\alpha}.\boldsymbol{\nabla}}{i} + \beta m + \beta M \left(\cos\theta(r) + i\gamma_5 \widehat{r}_a \tau_a \sin\theta(r)\right) \tag{12.43}$$

Proper-time regularization is used for the quark loop. The model parameters are fitted to vacuum values, as explained in section 5.5. The relation between the cut-off and the constituent quark mass is obtained by fitting the pion decay constant $f_\pi = 93$ MeV. The current quark mass m is fitted to the mass of the $\pi(138)$. The energy (12.18) is then minimized with respect to the shape of the profile function $\theta(r)$, subject to the boundary condition $\theta(r=0) = \pi$, $\theta(r \to \infty) = 0$. The results can be analyzed as a function of the constituent quark mass M. Table 12.1 lists some results obtained in [205].

The soliton solution represents a bound state of u and d quarks if $E_{soliton} < N_c M$. Table 12.1 shows that this is only achieved if the constituent quark mass M exceeds 414 MeV. Somewhat different values [174] are obtained with different regularization schemes. In the Nambu Jona-Lasinio model, an increase in the constituent quark mass M is equivalent to an increase in the coupling constant

Table 12.1 *Soliton data as a function of constituent quark mass M*

M (MeV)	372	414	465	558
$E_{soliton}$ (MeV)	1259	1247	1228	1190
$\frac{E_{soliton}}{N_c M}$	1.13	1.00	0.88	0.71
e_{val} (MeV)	197	174	146	97
E_{Dirac} (MeV)	669	725	790	900
$\sqrt{\langle r^2 \rangle}$ (fm)	0.76	0.69	0.64	0.57
$\sigma_N = m\frac{dE_{soliton}}{dm}$ (MeV)	54.2	50.0	45.9	38.3

g_s^2, as may be seen from the gap equation (5.14) and Table 5.1. Thus, a soliton is only formed when the coupling constant g_s^2 is strong enough to bind the quarks. This corresponds to a dimensionless quark–meson coupling constant (5.40) $g = \frac{M}{f_\pi} > 4.5$, which is very close to the critical coupling constant required to obtain solitons in the Gell-Mann Lévy model [169, 170]. Typical values used are shown in (11.21). The soliton energy varies very slowly with the constituent quark mass because the gain in energy due to the lowering of the valence 0^+ orbit, is compensated by the increase in the Dirac sea energy $E_{Dirac} = E_{soliton} - N_c e_v$. As a result the binding energy $E_{soliton} - N_c M$ does not exceed 20% of the mass: the soliton is a rather weakly bound state. Table 12.1 lists the isoscalar charge radius. If the constituent quark mass becomes too large, the soliton becomes too small: the more the soliton is bound, the smaller it gets. Table 12.1 also lists the nucleon sigma term, which is estimated experimentally to be $\sigma_N \approx 45$ MeV [130]. The soliton energy is of the order of 1200 MeV, which exceeds the nucleon mass $M_N = 939$ MeV by about 250 MeV. However, part of the soliton energy is due to spurious center-of-mass motion and rotational motion. Tentative estimates [205, 206] suggest that such corrections can lower the energy by 300 – 400 MeV.

12.9 Problems with soliton stability

In the present state of the art, soliton calculations are confronted with various instabilities. In the Diakonov Petrov model, described by the action (11.2), the chiral field is restricted to the chiral circle. This restriction is no longer imposed in the Nambu Jona-Lasinio model, described by an action such as (12.10). In principle, we should be able to use the Nambu Jona-Lasinio model to check whether a chiral field which remains close to the chiral circle is energetically favored over one which does not. Unfortunately, when hedgehog solitons are calculated with the Nambu Jona-Lasinio model and proper-time regularization, they are found to be unstable against collapse [153]. The instability occurs only when the chiral field is allowed to stray off the chiral circle, and it appears to be related to the proper-time regularization [154]. This is why solitons are calculated in the Diakonov Petrov model instead of the Nambu Jona-Lasinio model. Solitons can also be calculated when the Nambu Jona-Lasinio model is regularized using a 3-momentum cut-off, as in (6.14). When this is done, the instability against collapse disappears, but the Nambu Jona-Lasinio model fails to produce a bound state of quarks. When solitons are calculated in the renormalized Gell-Mann Lévy model, instabilities occur when the chiral fields acquires too high gradients [95]. This is related to the existence of low-lying Landau ghosts (see section 11.3.3).

12.10 A problem related to vector interactions

In section 4.3 we used the identity (4.14) to linearize the ($\mu = 0$) time components $\frac{g_v^2}{2}\left(q^\dagger \tau_a q\right)^2$ and $\frac{g_v^2}{2}\left(q^\dagger \gamma_5 \tau_a q\right)^2$ of the vector and axial interactions. This led to the non-hermitian Dirac hamiltonian (4.17). In the meson sector, the non-hermiticity of the Dirac hamiltonian is not a problem. Indeed, at the stationary point of the action, which represents the physical vacuum, the vector fields vanish. We calculate the meson propagators by expanding the action to second order in powers of the fields. The expansion of the quark determinant in powers of the non-hermitian operators iV_4 and $-i\gamma_5 A_4$ is straightforward. (In this section, we shall use the abridged notation $V_4 \equiv V_{4a}\tau_a$ for all the fields.)

In the soliton sector however, the time components ($\mu = 0$) of the vector fields V_4 and A_4 do not vanish. Furthermore, one can show that the stationary point of the action, at which the soliton occurs, is such that the time components of the vector fields acquire imaginary values $V_{40} = iV_0$ and $A_4 = iA_0$ [207]. At this stationary point, the Dirac hamiltonian becomes hermitian :

$$h = \frac{\boldsymbol{\alpha}.\boldsymbol{\nabla}}{i} + \beta m + \beta\left(S + i\gamma_5 P\right) + \boldsymbol{\alpha}.\boldsymbol{V} - V_0 - \gamma_5 \boldsymbol{\alpha}.\boldsymbol{A} + \gamma_5 A_0 \tag{12.44}$$

as might be expected for a Hartree approximation.

The hamiltonian (4.4) of the system, obtained by canonical quantization, is of course hermitian. The non-hermiticity of the Dirac hamiltonian is in fact a problem related to the regularization, as we now proceed to show. The occurrence of the imaginary auxiliary fields iV_4 and iA_4 in the Dirac hamiltonian (4.17) is due to the use of the identity (4.14) which leads to the linearization of the time component of the vector interaction. Had we used the identity (4.13) for all the interactions, we would have obtained the hermitian Dirac hamiltonian (12.44). However, the vector and scalar interactions $\frac{g_v^2}{2}\left(q^\dagger \tau_a q\right)^2$ and $-\frac{g_s^2}{2}\left(\bar{q}\tau_a q\right)^2$ appear with opposite signs in the hamiltonian (4.4). As a result, the use of the identity (4.13) for the vector interaction $\frac{g_v^2}{2}\left(q^\dagger \tau_a q\right)^2$ would lead to a diverging integral over the auxiliary fields.

We are however free to deform the integration contour of a field in such a way that the action should have a minimum as we pass through the stationary point. Such a deformation would allow us to use the identity (4.13) for all the interactions. However, the integration contour cannot be deformed when proper-time regularization is used for the quark determinant (see section 6.3.1), in spite of the fact that the regularized action is finite at the stationary point. Indeed, in section 12.4 , we have shown that the proper-time regularized action is a discontinuous and singular functional of the fields.

One suggested solution to this problem consists in modifying the proper-time regularization of the quark determinant [208]. Instead of using the expression (6.16), we use the following proper-time regularized quark determinant:

$$-Tr_\Lambda \ln D = \frac{1}{2}\int_{\frac{1}{\Lambda^2}}^{\infty} \frac{ds}{s} e^{-s\bar{D}D} + \frac{1}{2} Tr \ln \bar{D}/D \qquad D = \partial_\tau + h \qquad \bar{D} = -\partial_\tau + \bar{h} \tag{12.45}$$

where:

$$h = \frac{\boldsymbol{\alpha}.\boldsymbol{\nabla}}{i} + \beta m + \beta\left(S + i\gamma_5 P\right) + \boldsymbol{\alpha}.\boldsymbol{V} + iV_4 - \gamma_5 \boldsymbol{\alpha}.\boldsymbol{A} - i\gamma_5 A_4$$

$$\bar{h} = \frac{\boldsymbol{\alpha}.\boldsymbol{\nabla}}{i} + \beta m + \beta\left(S + i\gamma_5 P\right) + \boldsymbol{\alpha}.\boldsymbol{V} - iV_4 - \gamma_5 \boldsymbol{\alpha}.\boldsymbol{A} + i\gamma_5 A_4 \tag{12.46}$$

As long as V_4 and A_4 remain real, $\bar{D} = D^\dagger$ and the regularization (12.45) is the same as the proper-time regularization (6.16). However, $\bar{D} \neq D^\dagger$ at the stationary point where the fields acquire the imaginary values $V_4 = iV_0$ and $A_4 = iA_0$. The regularization (12.45) has several advantages. The operator $\bar{D}D$ is an analytic function of the fields, whereas $D^\dagger D$ is not. The first term of (12.45) is even in V_4 and V_0 whereas the second term is odd. This ensures that anomalous processes remain unregularized in the vacuum, where a stationary point exists with vanishing vector fields. The regularization (12.45) has also disadvantages. When V_4 and A_4 are imaginary (and V_0, A_0 real), the hamiltonians $h, \bar{h}$ and $h\bar{h}$ don't commute and the Dirac operators D and $\bar{D}$ cannot be simultaneously diagonalized. This greatly complicates soliton calculations. Last but not least, the operator $\bar{D}D$ is only positive definite for sufficiently weak vector fields. Whether negative eigenvalues of $\bar{D}D$ occur or not depends on the strength of the vector fields at the stationary point, and therefore on the model parameters. Negative eigenvalues of $\bar{D}D$ are found to occur dangerously close to the parameters required to form a hedgehog soliton [208].

Another suggested solution consists in using the regularization (12.45) and in expanding the action to second order only in the vector fields. Solitons are then calculated with this truncated action [209]. However, the validity of the expansion will obviously break-down in the vicinity of the region where the operator $\bar{D}D$ develops a negative eigenvalue. This would not be felt by a second order expansion in powers of the field. The problems encountered above are essentially due to the separation of the action into real and imaginary parts which are not regularized in the same way. The reason for proceeding this way is phenomenological. The imaginary part gives rise to anomalous processes which are known to fit experiment when left unregularized. As discussed in section 9.5, non-local regularizations of both the real and imaginary parts also leave anomalous processes unregularized. However, soliton calculations are considerably complicated in non-local regularizations. No results appear to have been published so far.

13

ROTATIONS OF SOLITONS IN FLAVOR SPACE

The hedgehog shape (12.40) of the chiral field is not invariant with respect to rotations generated by the angular momentum operator $\boldsymbol{j}$ nor by rotations generated by the isospin operator $\boldsymbol{t} = \frac{1}{2}\boldsymbol{\tau}$. It is only invariant with respect to rotations generated by the grand spin operator $\boldsymbol{G} = \boldsymbol{j} + \boldsymbol{t}$. The question then arises as to the nature and the quantum numbers of the system which is described by the hedgehog soliton. Indeed, it does not have definite angular momentum and isospin, whereas the observed hadrons do. Similarly, the hedgehog solution, or for that matter any localized soliton solution, is not translationally invariant. It describes a system with neither definite momentum nor definite center-of-mass position. Angular momentum, isospin and translational invariance appear as spontaneously broken symmetries. Such broken symmetries are not observed, because hadrons are *finite systems* with finite inertia parameters. Instead, they give rise to observable collective motion.

The breakdown of symmetry in finite systems is familiar in nuclear and molecular physics. It appears because the system is calculated in the mean field approximation. Had we calculated the partition function exactly, we would have found that it describes systems with definite angular momentum, isospin and linear momentum. In order to describe hadrons with proper quantum numbers, it is not necessary to find exact solutions. It is however necessary to go beyond the mean field approximation. We must include effects which are next to leading order in N_c: not all the effects, only the physically relevant ones. For example, to leading order in N_c, the nucleon $N\,(939)$ and the delta $\Delta\,(1232)$ would be degenerate particles. The hedgehog soliton describes a state which is some linear superposition of the two. In this chapter, we describe how to disentangle them.

The underlying physical picture is that of a slowly rotating deformed state. Indeed, if we rotate a hedgehog solution, a degenerate solution is obtained. There is no restoring force opposing spatial rotations. The only energy acquired by the rotation is a kinetic energy, proportional to the squared angular momentum and inversely proportional to a moment of inertia. One expects then that the low-lying excited states of the system, those which dominate the partition function, should consist of a rotational band generated by the rotating soliton. However, further effects need to be taken into account. The angular momentum acquired by the rotating soliton is quantized and cannot therefore take arbitrarily small values. The rotation gives rise to Coriolis forces which deform the soliton. They tend to align the spins with the rotation axis. The system eventually acquires axial symmetry about the rotation axis, at which point the collective rotation ceases. A baryon is composed of N_c quarks. The aligned spins cannot produce an angular momentum higher than 3/2, because of the low value $N_c = 3$. One may wonder whether the rotational band, which starts with the $N\,(939)$ which has spin and isospin $I = T = 1/2$, does not terminate with the $\Delta\,(1232)$ which has $I = T = 3/2$. This would be the case if the quark spins in the $\Delta\,(1232)$ were aligned. However, the rotating pion field also contributes to the angular momentum. Soluble models have been formulated to study the alignment of quark spins in a rotating soliton. They show that the quark spins in the $\Delta\,(1232)$ align only if the radius of the soliton is larger than the equilibrium sizes usually obtained, and that only about 25% of the spin is carried by the quarks in the nucleon and in the delta [210].

When the u and d current quark masses are taken to be equal and when electromagnetic interactions are neglected, there is also no restoring force opposing rotations in isospin space. If, in addition, the current mass of the s quark was equal to the current masses of the u and d quarks, there would be no restoring force opposing rotations in the full flavor space. However, the strange quark has a current mass $m_s \approx 200$ MeV . It opposes rotations of the soliton in the strange direction. If the moment of inertia of the soliton, which is proportional to N_c, is large enough, it becomes energetically favored for the system to generate the baryon octet by performing collective rotations in flavor space, in spite of the current strange quark mass, which opposes the rotation. This is the picture originally proposed for the octet generated by the Skyrmion [211] and applied to the Nambu Jona-Lasinio model [185, 186, 205, 212]. If, on the contrary, the current mass m_s of the strange quark is large enough, the baryon octet is better described by vibrations of the soliton in the strange direction, as proposed in [213, 214]. On the basis of presently available calculations [186, 205] it is difficult to decide which is the better description of the baryon octet. Because N_c is small, rotations and vibrations are not completely orthogonal collective modes and the two descriptions may significantly overlap.

13.1 Introduction of collective coordinates

Consider a system described by a regulated euclidean action, which we write in the form (4.20):

$$I\,(S, P, ...) = -Tr_\Lambda \ln\left(\partial_\tau + h + \beta m\right) + \frac{1}{2g_s^2}\int d_4x\left(S_a^2 + P_a^2\right) + ... \tag{13.1}$$

with the notation $S \equiv S_a \tau_a$ and $P = P_a \tau_a$ and where:

$$h = \frac{\boldsymbol{\alpha}.\boldsymbol{\nabla}}{i} + \beta\left(S_a \tau_a + i\gamma_5 P_a \tau_a + ...\right) \tag{13.2}$$

Typically, proper-time regularization could be used for the fermion determinant, in which case the action would have the form (6.18).

Consider the following transformation of the fields $S, P, ...$:

$$\left(S_a\tau_a + i\gamma_5 P_a\tau_a + ...\right) \rightarrow A(\alpha)\left(S_a\tau_a + i\gamma_5 P_a\tau_a + ...\right) A^{-1}(\alpha) \tag{13.3}$$

We shall specialize below to the case where the operator $A(\alpha)$ performs rotations in isospin and configuration space, in which case the parameters α become the angles which parametrize the rotation (see appendix B). In the following we refer to the α as *collective coordinates*, because, a priori, the operator $A(\alpha)$ need not be a rotation. In (13.3), the collective coordinates are introduced as specific transformations of the fields.

The idea is to promote the collective coordinates α to the rank of dynamical variables. With the substitution (13.3), the action becomes a functional $I(S, P, ..., \alpha)$ of the collective coordinates α and the partition function becomes:

$$Tre^{-\beta H} = \int D(\alpha) D(S) D(P) ... e^{-I(S,P,...,\alpha)} \tag{13.4}$$

If we could perform the integral over the fields $S, P, ...$ exactly, the redundant degrees of freedom α would have no effect because they would correspond to a redefinition of the fields. In the other extreme, we could limit the fields $S, P, ...$ to their classical values, which make the action stationary, and consider the collective coordinates α to be the only dynamical degrees of freedom. In this case the partition function would be:

$$Tre^{-\beta H} = \int D(\alpha) e^{-I(S,P,...,\alpha)} \tag{13.5}$$

with the understanding that the fields $S, P, ...$ are written at a stationary point of the action, typically a hedgehog-shaped solution of the equations (12.16). This is what has been done in most applications so far. This procedure neglects possible distortions of the solitons due to the rotation. One could also consider integrating out the quadratic fluctuations of the fields $S, P, ...$ in the partition function (13.4), in which case one has to face the problem that the fluctuating parts of the dynamical fields $S, P, ...$ have some overlap with the collective coordinates α. This course does not seem to have been pursued so far. In the case of simple solitons, the problem is addressed in chapter 8 of [215].

13.2 Rotations of a hedgehog soliton

To study the rotations of the hedgehog solution (12.40), we consider the case where the operator A is a product of a $SU(2)_f$ or $SU(3)_f$ rotation $A(\alpha)$ generated by the flavor operators t_a and a $SU(2)$ rotation $B(\beta)$ generated by the angular momentum operators j_a. The collective coordinates α_a and β_a are the angles which parametrize the two rotations. They can (but need not) be chosen to be Euler angles, defined in section B.1. In the following, rotations generated by the angular momentum operators will be referred to as rotations in configuration space and those which are generated by flavor (or isospin) operators as rotations in flavor (or isospin) space. We shall further assume that the angles are functions $\alpha(\tau)$ and $\beta(\tau)$ of the time variable τ only. For the sake of simplicity and without great loss of generality, we assume that the action depends only on the chiral field $S + i\gamma_5 P$. The angles α and β are introduced as collective coordinates by the following substitution of the fields:

$$S + i\gamma_5 P \to A(\alpha)\, B(\beta)\,(S + i\gamma_5 P)\, B^{-1}(\beta)\, A^{-1}(\alpha) \tag{13.6}$$

Using the cyclic permutation of the trace, we can write:

$$-Tr_\Lambda \ln\left(\partial_\tau + \frac{\boldsymbol{\alpha}.\boldsymbol{\nabla}}{i} + \gamma^0 m + \gamma^0 AB\,(S + i\gamma_5 P)\, B^{-1}A^{-1}\right)$$

$$= -Tr_\Lambda \ln\left(\partial_\tau + h + A^{-1}(\partial_\tau A) + B^{-1}(\partial_\tau B) + \gamma^0 A^{-1} m A\right) \tag{13.7}$$

We use the notation γ^0 for the Dirac matrix β in order not to confuse it with the angles β. The current quark mass term m is assumed to have the diagonal form (4.2) in flavor space. We assume that the u and d current quark masses m_u and m_d are equal. The current quark mass term m is then invariant in an $SU(2)_f$ isospin rotation. But when the strange quark mass $m_s > m_u$, the mass term m does not commute with the $SU(3)_f$ flavor rotation operator A. The last term of the action (13.1) does not change in the transformation (13.6). The action becomes:

$$I(S, P, \alpha, \beta)$$

$$= -Tr_\Lambda \ln\left(\partial_\tau + h - i\omega_a t_a - i\Omega_a j_a + \gamma^0 A^{-1} m A\right) + \frac{1}{2g_s^2}\int d_4x\left(S_a^2 + P_a^2\right) \tag{13.8}$$

In writing the form (13.8), we have used the expression (B.3). The $\omega_a = C_{ba}(\alpha)\,\dot{\alpha}$ and $\Omega_a = C_{ba}(\beta)\,\dot{\beta}$ are linear in the angular velocities $\dot{\alpha} = d\alpha/d\tau$ and $\dot{\beta} = d\beta/d\tau$.

The current quark mass m can be written in the form (5.50):

$$m = m_0\tau_0 + m_8\tau_8 \qquad m_0 = \frac{m_u + m_d + m_s}{2} \qquad m_8 = \frac{m_u + m_d - 2m_s}{2\sqrt{3}} \tag{13.9}$$

with $m_u = m_d$. Using the expression (B.62), we have:

$$A^{-1}(\alpha)\, m A(\beta) = -\frac{1}{2\sqrt{3}}\,(m_s - m_u)\, D_{8a}^{(8)}(\alpha)\, t_a \tag{13.10}$$

where $t_a = \frac{1}{2}\tau_a$ are the flavor generators and where $D_{ab}^{(8)}(\beta)$ is a Wigner function (B.58) for the octet representation.

We might be tempted to use the form (5.3) of the action, in which the current quark mass term βm is shifted from the quark determinant to the potential term. Had we done that, the action (13.8) would have been:

$$I\,(S, P, \alpha, \beta) = -Tr_{\Lambda} \ln\left(\partial_{\tau} + h - i\omega_a t_a - i\Omega_a j_a\right) - \frac{1}{2g_s^2} tr_f A^{-1} m A S + ... \tag{13.11}$$

This would present the advantage of avoiding an expansion in powers of $m_s - m_u$ since the last term could be diagonalized exactly. However, the action (5.3) is derived by performing the shift $m + S \to S$. This makes no difference if we integrate over the field S or as long as we remember that the stationary point is the shifted stationary point. However, it can make a difference if we perform the transformation (13.6) before integrating out the field. If we used the form (5.3), we would miss terms occurring in (13.8) which are proportional $(m_s - m_u)\,\Omega_a$. They do not occur in (13.11).

It is not very useful to search for a stationary point (α, β) of the action and to integrate the quadratic fluctuations about this point. Indeed, the energy of the system is invariant under global (time-independent) rotations generated by the angular momentum and isospin operators. As a result there is no potential energy term to make the gaussian integrals over α and β converge.

This is another way to say that the rotation is a large amplitude collective motion. Even so, the rotation can be slow, in which case it makes sense to expand the action in powers of the angular velocities $\dot{\alpha}$ and $\dot{\beta}$, which are proportional to ω_a and Ω_a. Thus our strategy will consist in expanding the action to second order in ω_a and Ω_a. We will thus obtain a euclidean action which describes a rotor. Then, instead of integrating out the angles α and β, we will simply recognize the hamiltonian of the rotor, which we will diagonalize in given $SU\,(2)$ and $SU\,(3)$ representations, as explained in section B.5.

13.3 $SU\,(2)_f$ rotations of a $G = 0$ soliton

The reader who is not familiar with the dynamics of a rotor is advised to study the appendix B.

We consider the simplest case where the fields S and P retain their hedgehog values and where the angles $\alpha\,(\tau)$ and $\beta\,(\tau)$ are treated as the only dynamical variables of the system. If the hedgehog-shaped chiral field was invariant under the transformation (13.6), the action would be invariant and the integration (13.4) over the angles α and β would simply lead to an overall inessential factor in the partition function.

Consider the simple case where a hedgehog soliton is calculated with u and d quarks with equal current masses. We further assume that $A\,(\alpha)$ represents an $SU\,(2)_f$ *isospin* rotation. In this case the term proportional to $(m_s - m_u)$ is not present. We expand the action (13.8) to second order in the angular velocities:

$$I\,(\alpha, \beta) = \frac{1}{2}\omega_a I_{ab} \omega_b + \frac{1}{2}\Omega_a J_{ab} \Omega_b + \omega_a K_{ab} \Omega_b + i\omega_a u_a + i\Omega_a v_a \tag{13.12}$$

The coefficients are:

$$u_a = -\langle t_a \rangle = \frac{\delta I}{\delta i\omega_a} \qquad v_a = -\langle j_a \rangle = \frac{\delta I}{\delta i\Omega_a}$$

$$I_{ab} = \frac{\delta^2 I}{\delta\omega_a \delta\omega_b} \qquad J_{ab} = \frac{\delta^2 I}{\delta\Omega_a \delta\Omega_b}$$

$$K_{ab} = \frac{\delta^2 I}{\delta\omega_a \delta\Omega_b} \tag{13.13}$$

The 'probe' $-i\omega_a t_a$ is antihermitian, so that the expectation values $\langle t_a \rangle, \langle j_a \rangle, ...$ and $I_{ab}, J_{ab}, ...$ should be calculated using the expressions I^g and R^{gg} given in equations (12.30) and (12.26). Valence corrections may need to be added, as explained in section 12.5.

The hedgehog field has the form (12.1) and the Dirac hamiltonian has the form (12.3) with a mass term $\gamma^0 m$ which is invariant with respect to isospin rotations. The Dirac hamiltonian commutes therefore with the grand spin operator $\boldsymbol{G} = \boldsymbol{j} + \boldsymbol{t}$. Using relations such as $e^{i\pi G_1} j_2 e^{-i\pi G_1} = -j_2$ and $e^{i\pi G_1} t_2 e^{-i\pi G_1} = -t_2$ and the fact that the soliton is composed of grand spin closed shells (see section 12.6), we have:

$$\langle j_a \rangle = \langle t_a \rangle = 0 \qquad K_{ab} = -I_{ab} \qquad I_{ab} = J_{ab} = \delta_{ab}\theta \tag{13.14}$$

where θ is a moment of inertia. It is to be calculated using the expression of R^{ff} given in (12.26). An unregularized expression is given in (13.38).

The action (13.12) reduces to the form:

$$I(\alpha, \beta) = \frac{\theta}{2} (\omega_a - \Omega_a)^2$$

$$= \frac{\theta}{2} \begin{pmatrix} \dot{\alpha} & \dot{\beta} \end{pmatrix} \begin{pmatrix} C(\alpha)\widetilde{C}(\alpha) & -C(\alpha)\widetilde{C}(\beta) \\ -C(\beta)\widetilde{C}(\alpha) & C(\beta)\widetilde{C}(\beta) \end{pmatrix} \begin{pmatrix} \dot{\alpha} \\ \dot{\beta} \end{pmatrix} \tag{13.15}$$

where, according to (B.3):

$$\omega_a = \widetilde{C}_{ab}(\alpha)\,\dot{\alpha}_b \qquad \Omega_a = \widetilde{C}_{ab}(\beta)\,\dot{\beta}_b \tag{13.16}$$

The moment of inertia θ is proportional to N_c because it is the expectation value of a colorless single particle operator.

The euclidean action (13.15) is a homogeneous quadratic form of the time derivatives $\dot{\alpha} = \frac{d\alpha}{d\tau}$ and $\dot{\beta} = \frac{d\beta}{d\tau}$. If we compare (B.37) with (B.32), we see that the real time action of the system has exactly the same form (13.15), in which the time derivatives are $\dot{\alpha} = \frac{d\alpha}{dt}$ and $\dot{\beta} = \frac{d\beta}{dt}$. (The real time action is sometimes improperly referred to as the Minkowski action.) The conjugate momenta are

$$\begin{pmatrix} \pi \\ p \end{pmatrix} = \theta \begin{pmatrix} C(\alpha)\widetilde{C}(\alpha) & -C(\alpha)\widetilde{C}(\beta) \\ -C(\beta)\widetilde{C}(\alpha) & C(\beta)\widetilde{C}(\beta) \end{pmatrix} \begin{pmatrix} \dot{\alpha} \\ \dot{\beta} \end{pmatrix} \tag{13.17}$$

and the hamiltonian is:

$$H_{coll} = \dot{\alpha}_a \frac{\partial I(\alpha,\beta)}{\partial \dot{\alpha}} + \dot{\beta} \frac{\partial I(\alpha,\beta)}{\partial \dot{\beta}} - I(\alpha,\beta) \tag{13.18}$$

$$= \frac{\theta}{2} \begin{pmatrix} \dot{\alpha} & \dot{\beta} \end{pmatrix} \begin{pmatrix} C(\alpha)\widetilde{C}(\alpha) & -C(\alpha)\widetilde{C}(\beta) \\ -C(\beta)\widetilde{C}(\alpha) & C(\beta)\widetilde{C}(\beta) \end{pmatrix} \begin{pmatrix} \dot{\alpha} \\ \dot{\beta} \end{pmatrix} \tag{13.19}$$

If we try to invert the matrix in (13.17) in order to express the momenta π and p in terms of the velocities $\dot{\alpha}$ and $\dot{\beta}$, we find that it can't be done. Instead, we discover the relation:

$$\begin{pmatrix} \widetilde{C}^{-1}(\alpha)C^{-1}(\alpha) & \widetilde{C}^{-1}(\alpha)C^{-1}(\beta) \\ \widetilde{C}^{-1}(\beta)C^{-1}(\alpha) & \widetilde{C}^{-1}(\beta)C^{-1}(\beta) \end{pmatrix} \begin{pmatrix} \pi \\ p \end{pmatrix} = 0 \tag{13.20}$$

The momenta π and p are therefore not independent.

Let us define intrinsic isospin and angular momentum operators $\mathcal{T}_a$ and $\mathcal{L}_a$ according to (B.12):

$$\mathcal{T}_a = C_{ab}^{-1}(\alpha)\pi_b = \theta\left(\widetilde{C}_{ab}(\alpha)\,\dot{\alpha}_b - \widetilde{C}_{ab}(\beta)\,\dot{\beta}_b\right)$$

$$\mathcal{L}_a = C_{ab}^{-1}(\beta)p_b = \theta\left(\widetilde{C}_{ab}(\beta)\,\dot{\beta}_b - \widetilde{C}_{ab}(\alpha)\,\dot{\alpha}_b\right) \tag{13.21}$$

These equations show that:

$$\mathcal{T} + \mathcal{L} = 0 \tag{13.22}$$

We also find that $\mathcal{T}_a - \mathcal{L}_a = 2\theta\left(\widetilde{C}_{ab}(\alpha)\,\dot{\alpha}_b - \widetilde{C}_{ab}(\beta)\,\dot{\beta}_b\right)$, so that the hamiltonian (13.19) is:

$$H_{coll} = \frac{1}{8\theta}\left(\mathcal{L}_a - \mathcal{T}_a\right)^2 \tag{13.23}$$

Upon canonical quantization, the intrinsic isospin and angular momentum operators $\mathcal{T}$ and $\mathcal{L}$ become differential operators acting respectively on the angles α and β. They obey the commutation rules (B.15):

$$[\mathcal{L}_a, \mathcal{L}_b] = -i\varepsilon_{abc}\mathcal{L}_c \qquad [\mathcal{L}_a, \mathcal{T}_b] = 0 \qquad [\mathcal{T}_a, \mathcal{T}_b] = -i\varepsilon_{abc}\mathcal{T}_c \tag{13.24}$$

By introducing the collective variables α and β, we have thus reduced the problem to the diagonalization of the hamiltonian (13.23), subject to the constraint (13.22). The constraint is due to the invariance of the hedgehog soliton with respect to rotations generated by the grand spin $\boldsymbol{G}$. The eigenfunctions of H_{coll} can be expressed in terms of Wigner functions (B.42). In section B.4 we show that the intrinsic isospin $\mathcal{T}_a$ acts on a Wigner function $D^{T*}_{M_T K_T}(\alpha)$ in exactly the same way as an angular momentum operator acts on the left on the state $\langle T, K_T|$. The intrinsic angular momentum $\mathcal{L}_a$ operator acts in the same way on the Wigner function $D^{J*}_{MK}(\beta)$. We can therefore satisfy the constraint (13.22) by coupling the two states $\langle TK_T|$ and $\langle JK|$ to zero total grand spin with a Clebsch–Gordan coefficient $\begin{bmatrix} T & J & 0 \\ K_T & K & 0 \end{bmatrix} = \delta_{JT}\delta_{KK_T}\frac{(-)^{T-K_T}}{\sqrt{2T+1}}$. The resulting normalized wavefunction is well known (see section 4.2g of [216]):

$$\psi_{T,J,M_T,M}(\alpha,\beta) = \sqrt{\frac{2T+1}{8\pi^2}}\sqrt{\frac{2J+1}{8\pi^2}}\sum_{K,K_T}\begin{bmatrix} T & J & 0 \\ K_T & K & 0 \end{bmatrix} D^{T*}_{M_T K_T}(\alpha)\, D^{J*}_{MK}(\beta)$$

$$= \delta_{JT}\frac{\sqrt{2T+1}}{8\pi^2}\sum_K (-)^{T-K}\, D^{T*}_{M_T,K}(\alpha)\, D^{J=T*}_{M,-K}(\beta) \qquad (13.25)$$

The wavefunction (13.25) is an eigenstate of the intrinsic grand spin $\mathcal{L}+\mathcal{T}$ belonging to a zero eigenvalue, so that the constraint (13.22) is satisfied. The hamiltonian (13.19) acting on the wavefunction has therefore the same eigenvalues are the hamiltonian $H = \frac{\mathcal{T}^2}{2\theta}$. Thus:

$$(\mathcal{L}+\mathcal{T})\,\psi_{T,J,M_T,M}(\alpha,\beta) = 0$$

$$H\psi_{T,J,M_T,M}(\alpha,\beta) = \frac{T(T+1)}{2\theta}\psi_{T,J,M_T,M}(\alpha,\beta) \qquad (13.26)$$

We see how the constraint (13.22) leads to $J = T$ states with equal spin and isospin. This in turn reflects the symmetry of the soliton with respect to rotations generated by the grand spin $\boldsymbol{G} = \boldsymbol{j}+\boldsymbol{t}$. The moment of inertia θ determines the mass difference $M_\Delta - M_N$ between the $\Delta(1232)$ and the $N(939)$. Values are given in Table 13.1.

The wavefunction (13.25) depends on two independent sets of Euler angles α and β because the orientations of the configuration and isospin space reference frames are independent and unrelated in the laboratory. The reader will have noticed however that, if the hedgehog field $S+i\gamma_5 P$ commutes with the grand spin operator $\boldsymbol{G} = \boldsymbol{j}+\boldsymbol{t}$, then:

$$A(\alpha)\,B(\beta)\,(S+i\gamma_5 P)\,B^{-1}(\beta)A^{-1}(\alpha) = A(\alpha)A^{-1}(\beta)\,(S+i\gamma_5 P)\,A(\beta)A^{-1}(\alpha)$$

$$= A(\gamma)\,(S+i\gamma_5 P)\,A^{-1}(\gamma) \qquad A(\gamma) = A(\alpha)\,A^{-1}(\beta) \qquad (13.27)$$

In other words, a rotation $B(\beta)$, generated by the angular momentum operators $\boldsymbol{j}$, is equivalent to an inverse rotation $A^{-1}(\beta)$ generated by the isospin operators

$\boldsymbol{t}$. The product of the isospin and configuration space rotations can therefore be represented by a single isospin rotation $A(\gamma)$. This is only true for fields which are invariant with respect to rotations generated by the grand spin operator $\boldsymbol{G}$. The substitution (13.6) is therefore equivalent to the substitution:

$$S + i\gamma_5 P \rightarrow A(\gamma)(S + i\gamma_5 P) A^{-1}(\gamma) \tag{13.28}$$

And indeed the wavefunction (13.25) can be represented in terms of the angle γ. We have $D^{T*}_{M,-K}(\beta) = (-)^{K-M} D^{T}_{-M,K}(\beta)$, so that:

$$\sum_K (-)^{T-K} D^{T*}_{M_T,K}(\alpha) D^{J=T*}_{M,-K}(\beta)$$

$$= (-)^{T-M} \sum_K \langle T, -M |A(\beta)| T, K\rangle \langle T, K \left|A^{-1}(\alpha)\right| T, M_T\rangle = (-)^{T-M} D^{T*}_{M_T,M}(\gamma) \tag{13.29}$$

where we used (13.27). The wavefunction (13.25) can therefore be written in the normalized form:

$$\psi_{J,M,T,M_T,M}(\gamma) = \sqrt{\frac{2T+1}{8\pi^2}} (-)^{T-M} D^{J=T*}_{M_T,-M}(\gamma) \tag{13.30}$$

This is the form which is usually found in the literature. Note the normalization (B.43) of the Wigner functions. The isospin operators act on the wavefunction (13.30) as the lab frame operators $\mathcal{R}_a$ and the angular momentum operators act on it as the intrinsic operators $\mathcal{L}_a$ (see section B.2). Lab frame and intrinsic operators commute, as angular momentum and isospin operators do.

Had we used the substitution (13.28), we would have expanded the action to second order in the angular velocities $\dot{\gamma}$ and we would have obtained the hamiltonian

$$H_{coll} = \frac{1}{2\theta} \mathcal{T}_a^2 \qquad \mathcal{T}_a = C^{-1}_{ab}(\gamma) p_b = \frac{1}{i} C^{-1}_{ab}(\gamma) \frac{\partial}{\partial \gamma_b} \tag{13.31}$$

The spectrum is again $E_T = T(T+1)/2\theta$ as in (13.26). Thus, for $G = 0$ solitons, it is sufficient to introduce collective coordinates corresponding to isospin rotations. In the next section, we show that the isospin T is even or odd, depending on whether the soliton is composed of an even or odd number of quarks.

13.4 The coupling of operators to the rotating soliton

To calculate the expectation value of a quark operator $q^\dagger f q$, we can probe the system by adding the operator f to the Dirac hamiltonian: $h \rightarrow h + f$. The operator f could also represent an electroweak current to which the quarks couple and which is treated perturbatively. We then expand the action to first order in

f, as done in section 12.5. When collective coordinates are introduced, a coupling occurs between the operator f, acting on the quark variables, and the collective rotational motion. Let us introduce collective coordinates γ corresponding to isospin rotations as in (13.28). We assume that, under rotations generated by the isospin operators, the operator f is a tensor f_a^k of rank k and component a. If not, it can always be expanded on the basis of such operators. We then have:

$$A^{-1}(\gamma)\, f_a^k A(\gamma) = D_{ab}^{k*}(\gamma)\, f_b^k \tag{13.32}$$

For example, if the operator f were proportional to t_a, it would be the component a of a tensor of rank $k = 1$ (a vector). After introducing the collective coordinates γ and in the presence of the probe f_a^k, the action is:

$$I(S,P,\gamma) = -Tr_\Lambda \ln\left(\partial_\tau + h - i\Omega_a t_a + D_{ab}^{k*}(\gamma)\, f_b^k\right) + \frac{1}{2g_s^2}\int d_4x \left(S_a^2 + P_a^2\right) \tag{13.33}$$

where $\Omega_a = C_{ba}(\gamma)\,\dot{\gamma}_a$.

Since the operator $-i\Omega_a t_a$ is antihermitian and $A^{-1}(\gamma)\, f_a^k A(\gamma)$ usually hermitian, the second order cross-term arises from the unregularized imaginary part of the action (see section 12.5). It is therefore instructive to consider the expansion of the action in the limit $\Lambda \to \infty$ of infinite cut-off. The procedure is easily generalized to $SU(3)_f$ rotations and to independent rotations in flavor and configuration space.

Let us assume again that the fields keep their values at a stationary point of the action $I(S,P)$, calculated before the introduction of the probe and of the collective coordinates. We obtain an action for the collective coordinates γ by expanding the action to second order in the time derivatives $\dot{\gamma}$. We thus expand the action to second order in Ω and also to first order in f in order to probe the system. It is simple to derive the general formula:

$$-Tr\ln(\partial_\tau + h + v) = -Tr\ln(\partial_\tau + h) + \beta\sum_h \langle h|v|h\rangle - \beta\sum_{ph}\frac{\langle h|v|p\rangle\langle p|v|h\rangle}{e_p - e_h} + \ldots \tag{13.34}$$

where the particle and hole states are defined in (12.27). Using the perturbation $v = -i\Omega_a t_a + D_{ab}^{k*}(\gamma)\, f_b^k$, the second order expansion of the action is, to first order in f:

$$I(S,P,\gamma) = I(S,P) - i\beta\Omega_a\langle t_a\rangle + \beta\Omega_a\langle t_a t_b\rangle_c\Omega_b + \beta D_{ab}^{k*}(\gamma)\left\langle f_b^k\right\rangle$$

$$+ i\beta\Omega_c\left\langle t_c f_b^k\right\rangle_c D_{ab}^{k*}(\gamma) + i\beta D_{ab}^{k*}(\gamma)\left\langle f_b^k t_c\right\rangle_c\Omega_c \tag{13.35}$$

where, for any two operators f and f', the connected expectation values $\langle f\rangle$ and $\langle ff'\rangle_c$ are defined to be:

$$\langle f \rangle = \sum_h \langle h | f | h \rangle \qquad \langle f f' \rangle_c = \sum_{ph} \frac{\langle h | f | p \rangle \langle p | f' | h \rangle}{e_p - e_h} \tag{13.36}$$

The factors β arise from the final integral $\int d\tau$ over time, so that we recognize that the action (13.35) has the form (B.37) of a euclidean action of a rotor. The corresponding hamiltonian is given by the expression (B.34):

$$H_{coll} = \frac{1}{2} \left(\mathcal{T}_a + \langle t_a \rangle \right) \theta_{ab}^{-1} \left(\mathcal{T}_a + \langle t_a \rangle \right)$$

$$+ D_{ab}^{k*}(\gamma) \left\langle f_b^k \right\rangle_c + \Omega_c \left\langle t_c f_b^k \right\rangle_c D_{ab}^{k*}(\gamma) + D_{ab}^{k*}(\gamma) \left\langle f_b^k t_c \right\rangle_c \Omega_c \tag{13.37}$$

where the moment of inertia θ_{ab} is given by the Inglis formula:

$$\theta_{ab} = 2 \sum_{ph} \frac{\langle h | t_a | p \rangle \langle p | t_b | h \rangle}{e_p - e_h} \tag{13.38}$$

and where $\mathcal{T}_a$ is the intrinsic isospin operator defined in (13.31).

The collective hamiltonian (13.37) is obtained by first quantizing the system in the absence of the perturbation f. In this quantization, the angular frequency Ω_a is represented by the quantized operator:

$$\Omega_a = \theta_{ab}^{-1} \left(\mathcal{T}_b + \langle t_b \rangle \right) \tag{13.39}$$

Thus, the last two terms of (13.37) are:

$$\Omega_c \left\langle t_c f_b^k \right\rangle_c D_{ab}^{k*}(\gamma) + D_{ab}^{k*}(\gamma) \left\langle f_b^k t_c \right\rangle_c \Omega_c$$

$$= \theta_{cs}^{-1} \left(\mathcal{T}_s + \langle t_s \rangle \right) \left\langle t_c f_b^k \right\rangle_c D_{ab}^{k*}(\gamma) + D_{ab}^{k*}(\gamma) \left\langle f_b^k t_c \right\rangle_c \theta_{cs}^{-1} \left(\mathcal{T}_s + \langle t_s \rangle \right) \tag{13.40}$$

The order in which the quantized operator Ω_c is written relative to the Wigner function $D_{ab}^{k*}(\gamma)$ is important and has been a subject of controversy [217, 218]. The path integral formulation does not specify the order in which the operators should appear. We have adopted the order in which they appear in time-independent perturbation theory. It agrees with the one which is used in [185] for example.

The second term of the collective hamiltonian (13.37) allows one to calculate expectation values and transition matrix elements of the operator f between the rotational states (13.30). The last two terms are perturbations which express the coupling between the operator f and the rotational motion. They lead to corrections in expectation values of operators (such as the axial constant g_A) due to the rotation of the soliton. Matrix elements of Wigner functions are given by (B.50) in terms of Clebsch–Gordan coefficients.

Table 13.1 *Proton and neutron properties*

M (MeV)	370	420	450	Exp.
$\langle r^2 \rangle_p$ (fm^2)	0.85	0.70	0.66	0.74
$\langle r^2 \rangle_n$ (fm^2)	-0.22	-0.18	-0.18	-0.11
μ_p (n.m.)	2.12	2.03	1.86	1.79
μ_n (n.m.)	-1.44	-1.41	-1.29	-1.91
$\left\lvert \frac{\mu_p}{\mu_n} \right\rvert$	1.47	1.44	1.44	1.46
$M_\Delta - M_N$ (MeV)	213	280	314	293
g_A	1.26	1.21	1.13	1.26
$g_{\pi NN}\,(q=0)$	10.2	10.3	10.3	13.1

There are alternative ways of deriving a collective hamiltonian such as (13.37). We could have performed first and second order perturbation theory in the Fock space of the combined collective hamiltonian H_{coll} and of the intrinsic hamiltonian $H_{int} = \int d_3xq^\dagger hq$. This would have added to the denominators $e_p - e_h$, which are the excitation energies of H_{int}, the excitation energies of H_{coll}. These are however assumed smaller and down by a factor of N_c. We could have also included the first order modifications of the fields $S, P, ...$ caused by the probe.

If we want to know whether the isospin of the system is even or odd, we can probe the system with the operator $e^{2i\pi t}$, where t is the component of the isospin operator around any given axis. This introduces the probe $A^{-1}(\gamma)\, e^{2i\pi t} A(\gamma) = e^{2i\pi t'}$ to the collective hamiltonian (13.37), where t' is the component of the isospin operator about the rotated axis. The expectation value of the probe in the rotating system is $\left\langle e^{2i\pi t'} \right\rangle = (-)^N$, where N is the number of u and d quarks which compose the soliton.

13.5 Calculated nucleon properties

Calculated baryon properties are reviewed in [185, 186]. They are derived from an $SU(2)_f$ hedgehog soliton, calculated with a proper-time regulated Diakonov Petrov model (see section 12.8), including only u and d quarks. As explained in section 5.5, the current quark mass m is determined by fitting the pion mass $m_\pi = 139$ MeV and, for a given constituent quark mass M, the cut-off can be determined by fitting the pion decay constant $f_\pi = 93$ MeV. Baryon properties can then be calculated for different values of the constituent quark mass M. Table 13.1, taken from [185], lists some proton and neutron properties.

The calculated axial constant g_A undergoes an approximately 40% increase due to the coupling of the axial charge operator with the rotational motion of the soliton [62, 218]. The calculated nucleon mass is discussed in section 12.8.

13.6 $SU(3)_f$ rotations of the $G = 0$ soliton

$SU(3)_f$ rotations of the soliton have been calculated in the framework of the Diakonov Petrov model (section 11.1), including u, d and s quarks. The euclidean action has the form:

$$I(\theta) = -Tr_\Lambda \ln(\partial_\tau + h + \beta m) \qquad h = \frac{\boldsymbol{\alpha}.\boldsymbol{\nabla}}{i} + \beta M U \qquad U = e^{i\gamma_5\theta(r)\Sigma_{a=1}^{3} r_a\tau_a} \tag{13.41}$$

The chiral field, represented by the matrix U, has the form of an embedding of an $SU(2)_f$ hedgehog field in an $SU(3)_f$ matrix:

$$U = e^{i\gamma_5\theta(r)\Sigma_{a=1}^{3} r_a\tau_a} = \begin{pmatrix} e^{i\gamma_5\theta(r)r_a\tau_a} & 0 \\ 0 & 1 \end{pmatrix} \tag{13.42}$$

and M is a constant, equal to the u and d constituent quark mass. The current quark mass matrix m has the diagonal form (4.2) with $m_s > m_u = m_d$. A static (non-rotating) soliton is formed by adding N_c quarks to the bound 0^+ orbit, displayed in Fig.12.1. The orbit is composed of only u and d quarks. When the chiral field has the form (13.42), the strange quarks are in their unperturbed vacuum orbits, so that the soliton is the same as the $SU(2)_f$ soliton, calculated with u and d quarks only and discussed in sections 12.8 and 13.3. In section 13.3, we discussed the $SU(2)_f$ rotations of the soliton generated by the three isospin operators $t_{a=1,2,3}$. In this section, we discuss the $SU(3)_f$ rotations of the soliton generated by the eight flavor generators $t_{a=1,\ldots,8}$. The strange quarks participate in these rotations. The $SU(3)_f$ rotations generate the baryon octet, which turns out to have admixtures of higher $SU(3)_f$ representations.

We proceed exactly as in the case of $SU(2)_f$ rotations discussed in section 13.3. The reader who is not familiar with the dynamics of an $SU(3)$ rotor is advised to study the appendix B. The matrix U is invariant with respect to rotations generated by the grand spin operator $G_a = j_a + t_a$ with $a = 1, 2, 3$, and with respect to rotations generated by t_8. We therefore have a relation analogous to (13.27):

$$A(\alpha) B(\beta)(S + i\gamma_5 P) B^{-1}(\beta) A^{-1}(\alpha) = A(\gamma)(S + i\gamma_5 P) A^{-1}(\gamma) \tag{13.43}$$

where:

$$A(\gamma) = A(\alpha) A^{-1}(\beta) \tag{13.44}$$

In these equations, $B(\beta)$ is an $SU(2)$ rotation generated by the three angular momentum operators j_a, and β_a are three Euler angles, while $A(\alpha)$ is a flavor rotation generated by the eight flavor operators $t_{a=1,\ldots,8}$, and α_a are eight Euler angles. The rotation $A^{-1}(\beta)$ is generated by the three flavor operators $t_{a=1,2,3}$ which have the same commutation rules as the angular momentum operators. It is therefore sufficient to introduce collective coordinates corresponding to flavor rotations by a substitution of the form (13.28), in which $A(\gamma)$ represents an

$SU(3)_f$ rotation generated by the eight flavor operators t_a. This leads to the euclidean action:

$$I(\theta,\gamma) = -Tr_\Lambda \ln\left(\partial_\tau + h - i\Omega_a t_a - \frac{1}{2\sqrt{3}}(m_s - m_u) D^{(8)}_{8a}(\alpha)\gamma^0 t_a\right) \tag{13.45}$$

where we used (13.10).

We wish to expand the action to second order in the angular velocities $\dot{\gamma}$ and therefore in Ω_a, and also to second order in $(m_s - m_u)$, because it is not clear which, if any, of the two is the smaller parameter. The expansion of the action is restricted by the following properties of the (non-rotating) soliton:

$$\langle t_a \rangle = 0 \quad if \quad a \neq 8$$

$$\langle t_8 \rangle = \frac{N_c}{2\sqrt{3}} \qquad \left\langle \gamma^0 t_8 \right\rangle = \frac{1}{2\sqrt{3}} \left\langle \bar{u}u + \bar{d}d \right\rangle \tag{13.46}$$

The expectation values of all the flavor operators except t_8 vanish because only t_8 is a scalar with respect to isospin rotations. Using the explicit expression for τ_8 given in (A.22), we have $q^\dagger t_8 q = \frac{1}{2} q^\dagger \tau_8 q = \frac{1}{2\sqrt{3}}\left(u^\dagger u + d^\dagger d\right) - \frac{1}{\sqrt{3}} s^\dagger s$. When a soliton is formed from the hedgehog-shaped chiral field (13.42), N_c quarks fill the bound 0^+ orbit displayed in Fig.12.1. The orbit is composed of u and d quarks and no s quarks. Therefore the number $\left\langle s^\dagger s\right\rangle$ of strange quarks in the (non-rotating) soliton is zero and the number $\left\langle u^\dagger u + d^\dagger d\right\rangle$ of u and d quarks is N_c. The value given for $\langle t_8 \rangle$ in (13.46) follows. The value given for $\left\langle \gamma^0 t_8 \right\rangle$ is obtained by the same argument. The strange quark condensate $\langle \bar{s}s \rangle$ does not contribute, because it has the same value as in the vacuum. The u and d condensates are often expressed in terms of the nucleon sigma term $\sigma = m\frac{\partial E_{soliton}}{\partial m} = \left\langle \bar{u}u + \bar{d}d \right\rangle$.

If we make use of the invariance of U with respect to rotations generated by G_1, G_2, G_3 and t_8, we find that the moment of inertia tensor reduces to the form:

$$\begin{aligned} &\theta_{ab} = A\delta_{AB} + B\delta_{\alpha\beta} + iC f_{8\alpha\beta} \qquad (A,B) = (1,2,3) \qquad (\alpha,\beta) = (4,5,6,7) \\ &\theta_{88} = 0 \end{aligned} \tag{13.47}$$

With the restrictions (13.46) and (13.47), the expansion of the action (13.45) becomes:

$$I(\theta,\gamma) = -Tr_\Lambda \ln\left(\partial_\tau + h - i\Omega_a t_a - \frac{1}{2\sqrt{3}}(m_s - m_u) D^{(8)}_{8a}(\alpha)\gamma^0 t_a\right)$$

$$I(\theta) = I(\theta) - i\Omega_8 \frac{N_c}{2\sqrt{3}} + \frac{1}{2}\Omega_a \theta_{ab} \Omega_b$$

$$-\frac{1}{2\sqrt{3}}(m_s - m_u) D^{(8)}_{8a}(\alpha)\left\langle \gamma^0 t_a \right\rangle + ... \tag{13.48}$$

There are, in addition, terms which are quadratic in $(m_s - m_u)$ and cross-terms which are proportional to $(m_s - m_u)\Omega_a$. They are carefully spelled out in [185]

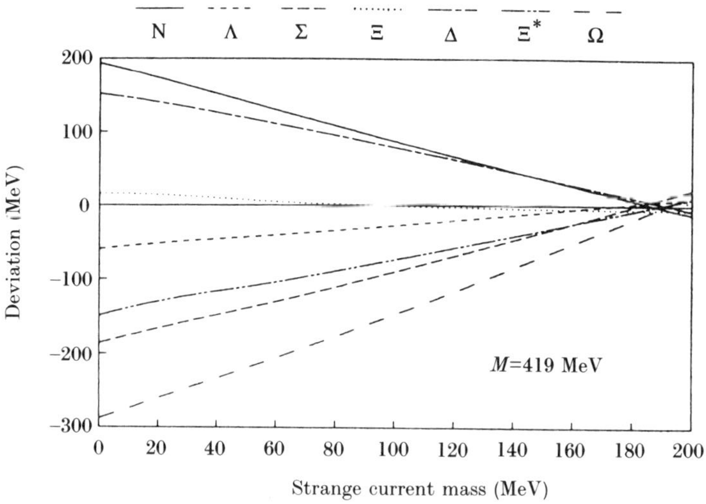

FIG. 13.1. The deviation of the calculated mass from the observed mass for the baryon octet, plotted as a function of the strange quark mass. From [205].

and need not be repeated here. The action (13.48) acquires the form of the euclidean action (B.37) of an $SU\,(3)$ rotor. The corresponding hamiltonian (B.34) can be diagonalized in a basis formed by the $SU\,(3)$ representations (B.55), using the method developed in [219] for the Skyrmion. The reader is referred to [185, 205] for further details. For the baryon octet, the $SU\,(3)$ rotor hamiltonian produces admixtures of higher representations.

13.7 Calculated properties of the $N\,(939)\,,\Lambda\,(1116)\,,\Sigma\,(11943)\,,\Xi\,(1318)$ octet

The mass spectrum can be calculated by diagonalizing the collective hamiltonian in all the $SU\,(3)_f$ multiplets. Figure (13.1), taken from [205], shows the deviation of the calculated and experimental masses, plotted as a function of the strange current quark mass m_s. A rotational zero point energy has been subtracted from the soliton energy. The calculated proton and neutron charge radii are $\langle r^2 \rangle_p = 0.78\ \text{fm}^2$ and $\langle r^2 \rangle_n = -0.09\ \text{fm}^2$. From Table 13.1, we see that they are closer to experiment than the values obtained in the $SU\,(2)_f$ soliton, which neglects the strange quark. The nucleon magnetic moments also agree better, as shown in Table 13.2, which lists the magnetic moments of the octet members [185].

Magnetic moments calculated in different models are discussed in [182]. Figure 13.2, taken from [185], shows the calculated proton and neutron electric and magnetic form factors. The calculated neutron electric form factor would

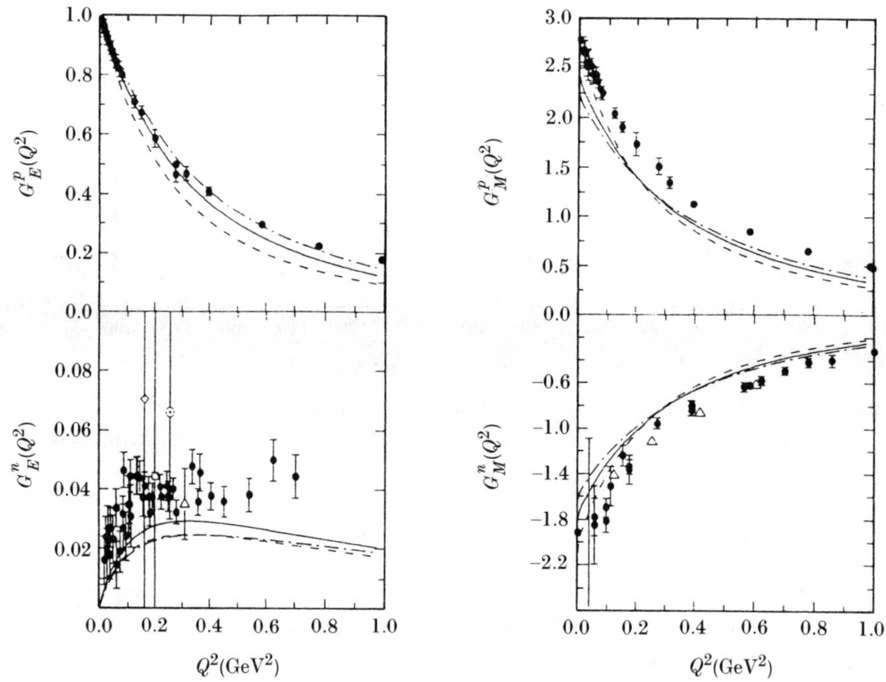

FIG. 13.2. Electric and magnetic form factors of the proton and the neutron. The full, dashed and dot-dashed lines are for constituent quark masses equal to 420, 370 and 450 MeV respectively. From [183].

Table 13.2 *Magnetic moments of the octet members. From [185].*

	μ (n.m.)	Exp.
p	2.39	2.79
n	-1.76	-1.91
Λ	-0.77	-0.61
Σ^+	2.42	2.46
Σ^0	0.75	
Σ^-	-0.92	-1.16
Ξ^0	-1.64	-1.25
Ξ^-	-0.68	-0.65

be about three times larger (and it would not fit the observed values) if only $SU\,(2)_f$ isospin rotations were taken into account.

Calculations of strange form factors and axial charges together with other spectroscopic properties are reviewed in [185]. In the review in [186], spectroscopic properties are also calculated in terms of soliton vibrations in the strange direction.

APPENDIX A

$SU(N)$ MATRICES, DIRAC MATRICES, FIERZ TRANSFORMATIONS

A.1 Properties of $SU(N)$ matrices

A useful review of the properties of the $SU(N)$ matrices can be found in [220, 221].

We denote by τ_a the N^2-1 hermitian traceless $N \times N$ matrices of an $SU(N)$ algebra. The matrices $\frac{1}{2}\tau_a$ are the generators of the $SU(N)$ group. The matrices are normalized to:

$$tr\tau_a\tau_b = 2\delta_{ab} \qquad (a,b) = 1,2,...,N^2-1 \tag{A.1}$$

For $N=2$, the matrices τ_a are the three Pauli matrices (A.21). For $N=3$, the matrices τ_a are the eight Gell-Mann matrices (A.22). The matrices τ_a and $i\tau_a$ span the space of complex $N \times N$ matrices, so that:

$$\tau_a\tau_b = \frac{2}{N}\delta_{ab} + (d_{abc} + if_{abc})\,\tau_c \tag{A.2}$$

and the following commutation relations follow:

$$[\tau_a, \tau_b] \equiv \tau_a\tau_b - \tau_b\tau_a = 2if_{abc}\tau_c \qquad \{\tau_a, \tau_b\} \equiv \tau_a\tau_b + \tau_b\tau_a = \frac{4}{N}\delta_{ab} + 2d_{abc}\tau_c$$

$$tr\,[\tau_a, \tau_b]\,\tau_c = 4if_{abc} \qquad tr\,\{\tau_a, \tau_b\}\,\tau_c = 4d_{abc} \tag{A.3}$$

Unless stated otherwise, a sum over repeated indices is assumed. The group structure functions f_{abc} and d_{abc} are respectively symmetric and antisymmetric. Then for $N=2$, we have $f_{abc} = \epsilon_{abc}$ and $d_{abc} = 0$. For $N=3$, the non-vanishing values are given in Table A.1 [7].

The following completeness relation is useful to derive Fierz identities:

$$\delta_{il}\delta_{kj} = \frac{1}{2}\left(\frac{2}{N}\delta_{ij}\delta_{kl} + (\tau_a)_{ij}\,(\tau_a)_{kl}\right) \tag{A.4}$$

Any $N \times N$ matrix can be expanded on the basis of the matrices τ_a :

$$K = a_0 + a_a\tau_a \qquad a_0 = \frac{1}{N}trK \qquad a_a = \frac{1}{2}tr\tau_aK \tag{A.5}$$

Table A.1 *Non-vanishing group structure functions for $N = 3$*

abc	f_{abc}	abc	d_{abc}	abc	d_{abc}
123	1	118	$1/\sqrt{3}$	355	$1/2$
147	$1/2$	146	$1/2$	366	$-1/2$
156	$-1/2$	157	$1/2$	377	$-1/2$
246	$1/2$	228	$1/\sqrt{3}$	448	$-1/(2\sqrt{3})$
257	$1/2$	247	$-1/2$	558	$-1/(2\sqrt{3})$
345	$1/2$	256	$1/2$	668	$-1/(2\sqrt{3})$
367	$-1/2$	338	$1/\sqrt{3}$	778	$-1/(2\sqrt{3})$
458	$\sqrt{3}/2$	344	$1/2$	888	$-1/\sqrt{3}$
678	$\sqrt{3}/2$				

For $SU(2)$ and $SU(3)$ the determinant of the matrix K is given by:

$$\det K = a_0^2 - a_a^2 \qquad (N = 2)$$

$$\det K = a_0^3 - a_0 a_a^2 + \frac{2}{3} d_{abc} a_a a_b a_c \qquad (N = 3) \tag{A.6}$$

If τ_a is a matrix representation of the $SU(N)$ group, then the transposed matrix $-\widetilde{\tau}_a$ belongs to the conjugate representation. In the case of $SU(2)$ only, we have $\widetilde{\tau}_a = -\tau_2 \tau_a \tau_2$. As a result the $SU(2)$ representations are self-conjugate. This is why ungauged euclidean actions involving $SU(2)$ chiral fields have vanishing imaginary parts.

A.1.1 *Fierz transformations of products of $SU(N)$ matrices*

A four-quark interaction has two $\bar{q}q$ channels and one qq channel. It is often expressed as a product of bilinears in one such channel, such as for example the vertex $(\bar{q}\gamma^\mu \lambda_a q)(\bar{q}\gamma^\mu \lambda_a q)$ of the gluon exchange interaction. The purpose of Fierz transformations is to express such products of bilinears in terms of products of bilinears in one of the other two channels. There are therefore two distinct Fierz transformations of the gluon exchange vertex: one which expresses it in the other $\bar{q}q$ channel and one which expresses it in the diquark qq channel.

The first Fierz transformation is obtained by writing $(\tau_a)_{ij}(\tau_b)_{kl} = (\tau_a)_{i\bar{j}}(\tau_b)_{k\bar{l}}\delta_{\bar{j}j}\delta_{l\bar{l}}$ and in using (A.4) to express the factor $\delta_{\bar{j}j}\delta_{l\bar{l}}$:

$$(\tau_a)_{ij}(\tau_b)_{kl} = (\tau_a)_{i\bar{j}}(\tau_b)_{k\bar{l}} \frac{1}{2}\left(\frac{2}{N}\delta_{\bar{j}l}\delta_{\bar{l}j} + (\tau_c)_{\bar{j}l}(\tau_c)_{\bar{l}j}\right)$$

$$= \frac{1}{2}\left(\frac{2}{N}(\tau_a)_{il}(\tau_b)_{kj} + (\tau_a\tau_c)_{il}(\tau_b\tau_c)_{kj}\right) \tag{A.7}$$

The second Fierz transformation is obtained by writing $(\tau_a)_{ij}(\tau_b)_{kl} = (\tau_a)_{i\bar{j}}(\tau_b)_{\bar{k}l}\delta_{j\bar{j}}\delta_{k\bar{k}}$ and in using the completeness relation (A.4) for the product $\delta_{j\bar{j}}\delta_{k\bar{k}} = \delta_{\bar{j}j}\delta_{k\bar{k}}$:

$$(\tau_a)_{ij}(\tau_b)_{kl} = \frac{1}{2}\left(\frac{2}{N}(\tau_a)_{ik}(\widetilde{\tau}_b)_{lj} + (\tau_a\tau_c)_{ik}(\widetilde{\tau}_b\tau_c)_{lj}\right) \tag{A.8}$$

The only difference between the Fierz transformations (A.7) and (A.8) is the replacement of $(\tau_b)_{lj}$ by $(\widetilde{\tau}_b)_{lj}$.

Let us apply this to the scalar operator $\tau_a\tau_a$. We use the multiplication law (A.2) to express (A.7) in the form:

$$(\tau_a)_{ij}(\tau_a)_{kl} =$$

$$= \frac{1}{N}(\tau_a)_{il}(\tau_a)_{kj} + \frac{2}{N^2}\left(N^2-1\right)\delta_{il}\delta_{kj} + \frac{1}{N}d_{aab}\left((\tau_b)_{il}\delta_{kj} + \delta_{il}(\tau_b)_{kj}\right)$$

$$+\frac{1}{2}d_{acb}d_{acd}(\tau_b)_{il}(\tau_d)_{kj} - \frac{1}{2}f_{acb}f_{acd}(\tau_b)_{il}(\tau_d)_{kj} \tag{A.9}$$

where we used $(\delta_{ac})_{il} = \delta_{ac}\delta_{il}$. We next use the identities:

$$d_{aac} = 0 \qquad d_{abc}d_{ab\rho} = \frac{N^2-4}{N}\delta_{c\rho} \qquad f_{abc}f_{ab\rho} = N\delta_{c\rho} \tag{A.10}$$

to reduce the first Fierz identity of a scalar to the form:

$$(\tau_a)_{ij}(\tau_a)_{kl} = \frac{2}{N^2}\left(N^2-1\right)\delta_{il}\delta_{kj} - \frac{1}{N}(\tau_a)_{il}(\tau_a)_{kj} \tag{A.11}$$

The repeated indices are summed from 1 to N^2-1.

To obtain the second Fierz transform, we separate the N^2-1 matrices τ_a into two classes: the $(N+2)(N-1)/2$ symmetric matrices $\tau^S = \widetilde{\tau}^S$ and the $N(N-1)/2$ antisymmetric matrices $\tau^A = -\widetilde{\tau}^A$. In the case of the $SU(2)$ Pauli matrices, they are:

$$\tau^S = (\tau_1, \tau_3) = (\widetilde{\tau}_1, \widetilde{\tau}_3) \qquad \tau^A = \tau_2 = -\widetilde{\tau}_2 \tag{A.12}$$

In the case of the $SU(3)$ Gell-Mann matrices, they are:

$$\tau^S = (\tau_1, \tau_3, \tau_4, \tau_6, \tau_8) = (\widetilde{\tau}_1, \widetilde{\tau}_3, \widetilde{\tau}_4, \widetilde{\tau}_6, \widetilde{\tau}_8)$$

$$\tau^A = (\tau_2, \tau_5, \tau_7) = (-\widetilde{\tau}_2, -\widetilde{\tau}_5, -\widetilde{\tau}_7) \tag{A.13}$$

The second Fierz transform (A.8) of a scalar can thus be written in the form:

$$(\tau_a)_{ij}(\tau_a)_{kl} = \frac{1}{N}(\tau_a)_{ik}(\tau_a)_{jl} + \frac{1}{2}\left(\tau^S\tau_c\right)_{ik}\left(\tau^S\tau_c\right)_{lj} - \frac{1}{2}\left(\tau^A\tau_c\right)_{ik}\left(\tau^A\tau_c\right)_{lj} \tag{A.14}$$

Again we use the multiplication law (A.2) for the products on the right hand side and we apply the identities:

$$d_{SSc} = 0 \qquad d_{AAc} = 0 \qquad \delta_{SS} - \delta_{AA} = N - 1$$

$$f_{Scb} f_{Scd} = f_{bSc} f_{dSc} = \frac{2}{N} \left(\delta_{bd} \delta_{SS} - \delta_{bS} \delta_{dS}\right) - d_{bSc} d_{dSc} \tag{A.15}$$

This yields the second Fierz identity

$$\left(\tau_a\right)_{ij} \left(\tau_a\right)_{kl} = \left(1 - \frac{1}{N}\right) \left(\frac{2}{N} \delta_{ik} \delta_{jl} + \left(\tau^S\right)_{ik} \left(\tau^S\right)_{jl}\right) + \left(1 + \frac{1}{N}\right) \left(\tau^A\right)_{ik} \left(\tau^A\right)_{jl} \tag{A.16}$$

A.2 $U\left(N\right)$ matrices

It is often useful to consider the N^2 matrices of a $U\left(N\right)$ algebra . We augment the set of the $N^2 - 1$ matrices τ_a by the following multiple of the unit matrix I:

$$\tau_0 = \sqrt{\frac{2}{N}} I \tag{A.17}$$

The following identities apply to the N^2 matrices of the $U\left(N\right)$ algebra:

$$\tau_a \tau_b = \left(d_{abc} + i f_{abc}\right) \tau_c$$

$$\left\{\tau_a, \tau_b\right\} = 2 d_{abc} \tau_c \qquad \left[\tau_a, \tau_b\right] = 2 i f_{abc} \tau_c \tag{A.18}$$

with:

$$d_{0ab} = \sqrt{\frac{2}{N}} \delta_{ab} \qquad f_{0ab} = 0 \tag{A.19}$$

as well as the relations:

$$tr \tau_a \tau_b = 2 \delta_{ab} \qquad \delta_{il} \delta_{kj} = \frac{1}{2} \sum_{a=0}^{N^2-1} \left(\tau_a\right)_{ij} \left(\tau_a\right)_{kl} \tag{A.20}$$

The four $U\left(2\right)$ matrices are:

$$\tau_0 = \begin{pmatrix} 1 & 0 \\ 0 & 1 \end{pmatrix} \quad \tau_1 = \begin{pmatrix} 0 & 1 \\ 1 & 0 \end{pmatrix} \quad \tau_2 = \begin{pmatrix} 0 & -i \\ i & 0 \end{pmatrix} \quad \tau_3 = \begin{pmatrix} 1 & 0 \\ 0 & -1 \end{pmatrix} \tag{A.21}$$

The three matrices $\tau_{a=1,2,3}$ are called the Pauli matrices.

The nine $U(3)$ matrices are:

$$\tau_0 = \begin{pmatrix} \sqrt{\frac{2}{3}} & 0 & 0 \\ 0 & \sqrt{\frac{2}{3}} & 0 \\ 0 & 0 & \sqrt{\frac{2}{3}} \end{pmatrix} \qquad \tau_8 = \begin{pmatrix} \frac{1}{\sqrt{3}} & 0 & 0 \\ 0 & \frac{1}{\sqrt{3}} & 0 \\ 0 & 0 & -\frac{2}{\sqrt{3}} \end{pmatrix}$$

$$\tau_1 = \begin{pmatrix} 0 & 1 & 0 \\ 1 & 0 & 0 \\ 0 & 0 & 0 \end{pmatrix} \quad \tau_2 = \begin{pmatrix} 0 & -i & 0 \\ i & 0 & 0 \\ 0 & 0 & 0 \end{pmatrix} \quad \tau_3 = \begin{pmatrix} 1 & 0 & 0 \\ 0 & -1 & 0 \\ 0 & 0 & 0 \end{pmatrix}$$

$$\tau_4 = \begin{pmatrix} 0 & 0 & 1 \\ 0 & 0 & 0 \\ 1 & 0 & 0 \end{pmatrix} \quad \tau_5 = \begin{pmatrix} 0 & 0 & -i \\ 0 & 0 & 0 \\ i & 0 & 0 \end{pmatrix} \quad \tau_6 = \begin{pmatrix} 0 & 0 & 0 \\ 0 & 0 & 1 \\ 0 & 1 & 0 \end{pmatrix} \quad \tau_7 = \begin{pmatrix} 0 & 0 & 0 \\ 0 & 0 & -i \\ 0 & i & 0 \end{pmatrix} \qquad (A.22)$$

The eight matrices $\tau_{a=1,2,\ldots,8}$ are called the Gell-Mann matrices [222] .

A.3 Fierz transformations of Dirac matrices.

We consider matrices Γ_i ($i = S, V, T, A, P$) defined as follows:

$$\Gamma_S = I, \quad \Gamma_V = \gamma_\mu, \quad \Gamma_T = \sigma_{\mu\nu} = \frac{i}{2}[\gamma_\mu, \gamma_\nu], \qquad \Gamma_A = \gamma_\mu\gamma_5, \quad \Gamma_P = i\gamma_5 \qquad (A.23)$$

The phases are chosen such that $\bar{q}\Gamma q$ is hermitian. The first Fierz transformation is derived in Ref.[223]:

$$\begin{pmatrix} \delta_{ij}\delta_{kl} \\ (\gamma^\mu)_{ij}(\gamma_\mu)_{kl} \\ (\sigma^{\mu\nu})_{ij}(\sigma_{\mu\nu})_{kl} \\ (\gamma^\mu\gamma_5)_{ij}(\gamma_\mu\gamma_5)_{kl} \\ (i\gamma_5)_{ij}(i\gamma_5)_{kl} \end{pmatrix} = \begin{pmatrix} \frac{1}{4} & \frac{1}{4} & \frac{1}{4} & -\frac{1}{4} & -\frac{1}{4} \\ 1 & -\frac{1}{2} & 0 & -\frac{1}{2} & 1 \\ \frac{3}{2} & 0 & -\frac{1}{2} & 0 & -\frac{3}{2} \\ -1 & -\frac{1}{2} & 0 & -\frac{1}{2} & -1 \\ -\frac{1}{4} & \frac{1}{4} & -\frac{1}{4} & -\frac{1}{4} & \frac{1}{4} \end{pmatrix} \begin{pmatrix} \delta_{il}\delta_{kj} \\ (\gamma^\mu)_{il}(\gamma_\mu)_{kj} \\ (\sigma^{\mu\nu})_{il}(\sigma_{\mu\nu})_{kj} \\ (\gamma^\mu\gamma_5)_{il}(\gamma_\mu\gamma_5)_{kj} \\ (i\gamma_5)_{il}(i\gamma_5)_{kj} \end{pmatrix} \qquad (A.24)$$

(The matrix (A.24) is given incorrectly in [5] and [224].)

We can write the transformation (A.24) in the form:

$$(\Gamma^A)_{ij} \cdot (\Gamma_A)_{kl} = \sum_B M_{AB} (\Gamma^B)_{il} \cdot (\Gamma_B)_{kj} \qquad (A.25)$$

The second Fierz transformation can be obtained from the first using an argument given in [117]. We use the charge conjugation matrix C which has the properties:

$$C = i\gamma^2\gamma^0 = -\widetilde{C} = -C^{-1} = -C^\dagger \qquad C^{-1}\gamma_\mu C = -\widetilde{\gamma_\mu} \tag{A.26}$$

from which we deduce the transformations of the operators Γ_A:

$$\widetilde{\Gamma_A} = s_A C^{-1}\Gamma_A C = -s_A C\Gamma_A C = -s_A C\Gamma_A C^*$$

$$s = diag\,(1, -1, -1, 1, 1) \tag{A.27}$$

We then write:

$$\left(\Gamma^A\right)_{ij} \cdot (\Gamma_A)_{kl} = \left(\Gamma^A\right)_{ij} \cdot \left(\widetilde{\Gamma_A}\right)_{lk} = -s_A \left(\Gamma^A\right)_{ij} .C_{ls} \left(\Gamma_A\right)_{st} C^*_{tk} \tag{A.28}$$

We can then use (A.25) to obtain the second Fierz transform:

$$(\Gamma_A)_{ij} \cdot (\Gamma_A)_{kl} = -s_A \sum_B M_{AB} \left(\Gamma_B C^*\right)_{ik} \cdot (C\Gamma_B)_{lj} \tag{A.29}$$

Explicitly, this reads:

$$\begin{pmatrix} \delta_{ij}\delta_{kl} \\ (\gamma^\mu)_{ij}\,(\gamma_\mu)_{kl} \\ (\sigma^{\mu\nu})_{ij}\,(\sigma_{\mu\nu})_{kl} \\ (\gamma^\mu\gamma_5)_{ij}\,(\gamma_\mu\gamma_5)_{kl} \\ (i\gamma_5)_{ij}\,(i\gamma_5)_{kl} \end{pmatrix} = \begin{pmatrix} -\frac{1}{4} & -\frac{1}{4} & -\frac{1}{4} & \frac{1}{4} & \frac{1}{4} \\ 1 & -\frac{1}{2} & 0 & -\frac{1}{2} & 1 \\ \frac{3}{2} & 0 & -\frac{1}{2} & 0 & -\frac{3}{2} \\ 1 & \frac{1}{2} & 0 & \frac{1}{2} & 1 \\ \frac{1}{4} & -\frac{1}{4} & \frac{1}{4} & \frac{1}{4} & -\frac{1}{4} \end{pmatrix} \begin{pmatrix} {C^*}_{ik}C_{lj} \\ (\gamma^\mu C^*)_{ik}\,(C\gamma_\mu)_{lj} \\ (\sigma^{\mu\nu}C^*)_{ik}\,(C\sigma_{\mu\nu})_{lj} \\ (\gamma^\mu\gamma_5 C^*)_{ik}\,(C\gamma_\mu\gamma_5)_{lj} \\ (i\gamma_5 C^*)_{ik}\,(iC\gamma_5)_{lj} \end{pmatrix} \tag{A.30}$$

APPENDIX B

THE $SU(N)$ ROTOR

B.1 The hamiltonian and the intrinsic generators

Let j_a be a set of $N^2 - 1$ operators which satisfy the commutation relations:[1]

$$[j_a, j_b] = i f_{abc} j_c \tag{B.1}$$

where the f_{abc} are the structure functions of the $SU(N)$ group. The expressions below follow from the commutation rules (B.1) alone. We consider operators $A(\alpha)$ which act in the same space as the j_a operators and which are expressed in terms of $N^2 - 1$ parameters α_a. When the operators j_a are hermitian, we may consider them as the generators of the $SU(N)$ group. We may then think of the operators $A(\alpha)$ as unitary operators which perform $SU(N)$ 'rotations', parametrized by angles α_a which could be Euler angles. Euler angles for $SU(2)$ and $SU(3)$ are defined in (B.5) and (B.8).

We define a matrix C_{ab}, which depends on the angles α_a and such that:

$$A^{-1} \frac{\partial A}{\partial \alpha_a} = -i C_{ab} j_b \tag{B.2}$$

If the angles $\alpha_a(t)$ depend on a time parameter t, we have:

$$A^{-1} \dot{A} = -i \omega_a j_a \qquad \omega_a = \widetilde{C}_{ab} \dot{\alpha}_b \tag{B.3}$$

where $\dot{\alpha} = d\alpha/dt$. The invariant group measure is:

$$d(\alpha) = \prod_a d\alpha_a \left|\det C\right| \tag{B.4}$$

The invariant group measure is only defined to within a factor and care must be taken to check which convention is used.

In the case of the $SU(2)$ group, $f_{abc} = \varepsilon_{abc}$ and the j_a obey angular momentum commutation rules. The matrix A can be written in terms of three Euler angles (see appendix CIV of [225]):

$$A = e^{-i\alpha j_3} e^{-i\beta j_2} e^{-i\gamma j_3} \qquad (0 < \alpha, \gamma < 2\pi) \qquad (0 < \beta < \pi) \tag{B.5}$$

[1] The author wishes to thank Michel Bauer for help in preparing this appendix.

From the angular momentum commutation rules we deduce the identities:

$$e^{i\alpha j_1} j_2 e^{-i\alpha j_1} = j_2 \cos\alpha - j_3 \sin\alpha \qquad e^{i\alpha j_2} j_1 e^{-i\alpha j_2} = j_1 \cos\alpha + j_3 \sin\alpha \tag{B.6}$$

which express rotations of vectors. We deduce an explicit form of the ω_a in terms of the Euler angles α_a and their time derivatives $\dot{\alpha}_a$:

$$\begin{pmatrix} \omega_1 \\ \omega_2 \\ \omega_3 \end{pmatrix} = \begin{pmatrix} -\sin\beta\cos\gamma & \sin\gamma & 0 \\ \sin\beta\sin\gamma & \cos\gamma & 0 \\ \cos\beta & 0 & 1 \end{pmatrix} \begin{pmatrix} \dot{\alpha} \\ \dot{\beta} \\ \dot{\gamma} \end{pmatrix} \tag{B.7}$$

The expression (B.7) displays an explicit form of the matrix $\widetilde{C}_{ab}$ in the case of $SU(2)$ when the α_a are chosen to be the Euler angles.

In the case of the $SU(3)$ group, the structure functions f_{abc} are listed in Table A.1. The operator $A(\alpha)$ can be written in terms of eight Euler angles [219]:

$$A = e^{-i\alpha j_3} e^{-i\beta j_2} e^{-i\gamma j_3} e^{-i\nu j_4} e^{-i(\rho/\sqrt{3}) j_8} e^{-i\alpha' j_3} e^{-i\beta' j_3} e^{-i\gamma' j_3}$$

$$(0 < \alpha, \gamma, \alpha', \gamma' < 2\pi) \quad (0 < \beta, \beta' < \pi) \quad (0 < \nu < \pi/2) \quad (0 < \rho < 3\pi) \tag{B.8}$$

The $SU(N)$ rotor is a system which is described by a lagrangian of the form:

$$L(\alpha, \dot{\alpha}) = \frac{1}{2}\omega_a M_{ab} \omega_b + b_a \omega_a - V(\alpha) \qquad \left(M = \tilde{M}\right) \tag{B.9}$$

The $\alpha_a(t)$ are the dynamical variables of the system. The conjugate momenta are:

$$\pi_a = \frac{\partial I}{\partial \dot{\alpha}_a} = \frac{\partial I}{\partial \omega_b} \frac{\partial \omega_b}{\partial \dot{\alpha}_a} = C_{ab}(M_{bc}\omega_c + b_b) \tag{B.10}$$

We can express the ω_a in terms of the π_a:

$$\omega_a = M_{ab}^{-1} C_{bc}^{-1} \pi_c - M_{ab}^{-1} b_b = M_{ab}^{-1}(\mathcal{L}_b - b_b) \tag{B.11}$$

where we defined the *intrinsic generators*:

$$\mathcal{L}_a = C_{ab}^{-1} \pi_b \tag{B.12}$$

If we express the hamiltonian $H = \dot{\alpha}_a p_a - L$ in terms of the dynamical variables α and their conjugate momenta π, we find that it can be written as a quadratic form of the intrinsic generators $\mathcal{L}_a$:

$$H(\pi, \alpha) = \frac{1}{2}(\mathcal{L}_a - b_a) M_{ab}^{-1} (\mathcal{L}_b - b_b) + V(\alpha) \tag{B.13}$$

We quantize the system by applying canonical quantization rules:

$$[\pi_a, \alpha_b] = \frac{1}{i}\delta_{ab} \qquad \pi_a = \frac{1}{i}\frac{\partial}{\partial \alpha_a} \qquad \mathcal{L}_a = \frac{1}{i}C^{-1}_{ab}\frac{\partial}{\partial \alpha_b} \tag{B.14}$$

The $\mathcal{L}$ become differential operators acting on the angles α. The precise form of the intrinsic generators $\mathcal{L}$ depends on the choice of coordinates α. Let us show however that the $\mathcal{L}$ obey commutation rules which depend only on the structure functions f_{abc} of the $SU(N)$ group and which are independent of the choice of coordinates α_a. We first note that $\mathcal{L}_a A = \frac{1}{i}C^{-1}_{ab}\frac{\partial A}{\partial \alpha_b}$. We multiply on the left by A^{-1} and use (B.2) to obtain:

$$A^{-1}(\mathcal{L}_a A) = -j_a \tag{B.15}$$

Let us calculate:

$$A^{-1}[\mathcal{L}_a, \mathcal{L}_b]A = [j_a, j_b] = if_{abc}j_c = -if_{abc}A^{-1}\mathcal{L}_c A \tag{B.16}$$

We deduce the commutation relation:

$$[\mathcal{L}_a, \mathcal{L}_b] = -if_{abc}\mathcal{L}_c \tag{B.17}$$

Instead of working with the N^2-1 operators j_a, it is often useful to work with the N^2 operators G_{ij} defined as follows:

$$G_{ij} = j_a(\tau_a)_{ji} \qquad j_a = \frac{1}{2}G_{ij}(\tau_a)_{ji} \tag{B.18}$$

In the expression (B.18), the τ_a are the N^2-1 traceless $SU(N)$ matrices, discussed in section (A.1). The N^2 operators G_{ij} are not independent, because $G_{ii} = 0$. The operators G_{ij} obey the commutation rules:

$$[G_{ik}, G_{jl}] = \delta_{il}G_{jk} - \delta_{jk}G_{il} \tag{B.19}$$

In the fundamental representation, we have $G_{ij} = |j\rangle\langle i| - \frac{1}{N}\delta_{ij}I$, where $I = |i\rangle\langle i|$ is the unit operator. The two operators:

$$C^{(2)} = G_{ij}G_{ji} = 2j_a^2 \qquad C^{(3)} = G_{ij}G_{jk}G_{kl} = 2d_{abc}j_a j_b j_c - Nj_a^2 \tag{B.20}$$

commute with the N^2-1 operators j_a. They are called Casimir operators. The anticommutation relation $\{\tau_a, \tau_b\} = \frac{4}{N}\delta_{ab} + 2d_{abc}\tau_c$ holds only for the $SU(N)$ matrices. No such rule exists in general for the operators j_a. It only holds for the operators $2j_a = \tau_a$ in the fundamental representation.

B.2 Intrinsic and lab frame generators

We define a matrix D_{ab}, which depends on the angles α_a and such that:

$$A\frac{\partial A^{-1}}{\partial \alpha_a} = iD_{ab}j_b \tag{B.21}$$

The matrix D_{ab} may be related to the matrix C_{ab} defined in (B.2) as follows:

$$A\frac{\partial A^{-1}}{\partial \alpha_a}A = -AA^{-1}\frac{\partial A}{\partial \alpha_a} = iD_{ab}j_bA \tag{B.22}$$

We multiply by A^{-1} on the left and we obtain the relation:

$$C_{ab}j_b = D_{ab}A^{-1}j_bA \tag{B.23}$$

We define 'lab frame' generators $\mathcal{R}_a$:

$$\mathcal{R}_a = D^{-1}_{ab}\pi_b = \frac{1}{i}D^{-1}_{ab}\frac{\partial}{\partial \alpha_b} \tag{B.24}$$

They are diffential operators, acting on the angles α. We note that $\mathcal{R}_aA^{-1} = \frac{1}{i}D^{-1}_{ab}\frac{\partial A^{-1}}{\partial \alpha_b}$. We multiply on the left by A and we obtain:

$$A\left(\mathcal{R}_aA^{-1}\right) = j_a \tag{B.25}$$

It follows that:

$$A\left[\mathcal{R}_a, \mathcal{R}_b\right]A^{-1} = \left[j_a, j_{bj}\right] = if_{abc}j_c = if_{abc}A\mathcal{R}_cA^{-1} \tag{B.26}$$

The lab frame generators therefore obey the commutation rules:

$$\left[\mathcal{R}_a, \mathcal{R}_b\right] = if_{abc}\mathcal{R}_c \tag{B.27}$$

We note the opposite signs of the commutation rules (B.17) and (B.27) of the intrinsic and lab frame generators. Let us further prove that the intrinsic and lab frame generators commute. We note that $(\mathcal{R}_aA) = \frac{1}{i}D^{-1}_{ab}\frac{\partial A}{\partial \alpha_b}A^{-1}A = -\frac{1}{i}D^{-1}_{ab}A\frac{\partial A^{-1}}{\partial \alpha_b}A$. It follows that:

$$(\mathcal{R}_aA) = -j_aA \tag{B.28}$$

From (B.15) and (B.28) we deduce that:

$$\mathcal{L}_a\mathcal{R}_bA = \mathcal{R}_b\mathcal{L}_aA = j_bAj_a \tag{B.29}$$

so that:

$$\left[\mathcal{L}_a, \mathcal{R}_b\right] = 0 \tag{B.30}$$

Two useful expressions are easily deduced from (B.15) and (B.25):

$$\left(\mathcal{L}_aA^{-1}\right) = j_aA^{-1} \qquad \left(\mathcal{R}_aA^{-1}\right) = A^{-1}j_a \tag{B.31}$$

It is important to distinguish the operators $\mathcal{L}_a$ and $\mathcal{R}_a$, which are differential operators acting in the space spanned by the angles α, from the operators j_a, A and G_{ij} which typically act in some Hilbert space.

B.3 The partition function

The lagrangian (B.9) of the system is a quadratic form of the time derivatives $\dot{\alpha}_a$:

$$L\left(\alpha, \dot{\alpha}\right) = \frac{1}{2}\,\dot{\alpha}_a\,N_{ab}\,\dot{\alpha}_b + v_a\,\dot{\alpha}_a - V\left(\alpha\right) \qquad \left(\dot{\alpha}_a \equiv \frac{d\alpha_a\left(t\right)}{dt}\right) \tag{B.32}$$

In the particular example (B.9), $N = CM\widetilde{C}$ and $v_a = C_{ab}b_b$. The momentum π_a, conjugate to α_a, is $\pi_a = \frac{\partial L}{\partial \dot{\alpha}_a} = N_{ab}\,\dot{\alpha}_b$. We express the hamiltonian in terms of the dynamical variables α_a and their conjugate momenta π_a:

$$H\left(\pi, \alpha\right) = \dot{\alpha}_a\,\frac{\partial L}{\partial\,\dot{\alpha}_a} - L = \frac{1}{2}\,\dot{\alpha}_a\,N_{ab}\,\dot{\alpha}_b + V\left(\alpha\right) \tag{B.33}$$

$$= \frac{1}{2}\left(\pi_a - v_a\right)N_{ab}^{-1}\left(\pi_b - v_b\right) + V\left(\alpha\right) \tag{B.34}$$

This hamiltonian is the same as the hamiltonian (B.13). It is simply written in terms of different variables. The first line of equation (B.34) is the classical energy of the system.

The path integral for the partition function is:

$$Tre^{-\beta H} = \int d\left(\pi\right)d\left(\alpha\right)e^{-\int_0^\beta d\tau\left(-i\pi_a\dot{\alpha}_a + H\left(\pi,\alpha\right)\right)} \qquad \left(\dot{\alpha}_a \equiv \frac{d\alpha_a\left(\tau\right)}{d\tau}\right) \tag{B.35}$$

The hamiltonian $H\left(\pi, \alpha\right)$ is a quadratic form of the momenta π_a. The gaussian integral over the momenta can be performed, with the result:

$$Tre^{-\beta H} = \int d\left(\alpha\right)e^{-\int_0^\beta d\tau L_{eucl}\left(\alpha,\dot{\alpha}\right)} \tag{B.36}$$

where the *euclidean* action is:

$$L_{eucl}\left(\alpha, \dot{\alpha}\right) = \frac{1}{2}\,\dot{\alpha}_a\,N_{ab}\,\dot{\alpha}_b + iv_a\,\dot{\alpha}_a + V\left(\alpha\right) \qquad \left(\dot{\alpha}_a \equiv \frac{d\alpha_a\left(\tau\right)}{d\tau}\right)$$

$$= \frac{1}{2}\omega_a M_{ab}\omega_b + i\omega_a b_a + V\left(\alpha\right) \tag{B.37}$$

Formally, the euclidean action (B.37) can be derived from the action (B.32) by an overall change in sign and by the substitution $t \to i\tau$. However, the partition function (B.35) is derived directly in terms of the hamiltonian $H\left(\pi, \alpha\right)$ without recourse to analytic continuation of the time variable.

In most applications there is no need to perform the path integral (B.36). When a euclidean action of the form (B.37) is obtained, it is simpler to recognize that the system is a rotor described either by the hamiltonian (B.13) or by the action (B.9). The hamiltonian (B.13) is then more efficiently diagonalized in a basis of states which form $SU\,(N)$ representations (Wigner functions), as described in the following sections.

B.4 The $SU(2)$ rotor

In the case of $SU(2)$, the three operators j_a obey angular momentum commutation rules $[j_a, j_b] = i\varepsilon_{abc} j_c$. The $SU(2)$ rotation group can be represented by states $|IM\rangle$. The quantum numbers I and M are eigenvalues of the commuting operators $\boldsymbol{j}^2 = j_1^2 + j_2^2 + j_3^2$ and j_3. The representation $\{I\}$ has a multiplicity $N = 2I+1$, corresponding to the $2I+1$ eigenvalues of j_3. The quantum number I corresponds to the eigenvalue $\boldsymbol{I}^2 |IM\rangle = I(I+1)|IM\rangle$ of the Casimir operator $\boldsymbol{I}^2$, which commutes with the three operators j_a. Matrix elements of the three operators j_a can be calculated from the relations:

$$j_3 |IM\rangle = M|IM\rangle \qquad (j_1 \pm i j_2)|IM\rangle = \sqrt{(I \mp M)(I \pm M + 1)}\,|IM \pm 1\rangle \tag{B.38}$$

The operator A is given by (B.5) and the matrix $\widetilde{C}_{ab}$ by the expression (B.7). The three Euler angles are $(\alpha_1, \alpha_2, \alpha_3) = (\alpha, \beta, \gamma)$. The generators $\mathcal{L}_a$ and $\mathcal{R}_a$ are respectively intrinsic and lab frame angular momentum operators. We can use (B.6) to obtain an explicit expression of the matrix D_{ab}, defined in (B.21):

$$D_{ab} = \begin{pmatrix} 0 & 0 & 1 \\ -\sin\alpha & \cos\alpha & 0 \\ \cos\alpha\sin\beta & \sin\alpha\sin\beta & \cos\beta \end{pmatrix} \tag{B.39}$$

The lab frame angular momentum operators may then be deduced from (B.24). We obtain the well-known expressions:

$$\mathcal{R}_\pm = \mathcal{R}_1 \pm i\mathcal{R}_2 = e^{\pm i\alpha}\left(i\cot\beta\left(\frac{\partial}{\partial\alpha}\right) \pm \left(\frac{\partial}{\partial\beta}\right) - i\frac{1}{\sin\beta}\left(\frac{\partial}{\partial\gamma}\right)\right) \qquad \mathcal{R}_3 = \left(\frac{\partial}{i\partial\alpha}\right) \tag{B.40}$$

The intrinsic angular momentum operators are similarly deduced from (B.7) and (B.14):

$$\mathcal{L}_\pm = \mathcal{L}_1 \pm \mathcal{L}_2 = e^{\mp i\gamma}\left(-i\cot\beta\left(\frac{\partial}{\partial\gamma}\right) \pm \left(\frac{\partial}{\partial\beta}\right) + \frac{i}{\sin\beta}\left(\frac{\partial}{\partial\alpha}\right)\right) \quad \mathcal{L}_3 = \left(\frac{\partial}{i\partial\gamma}\right) \tag{B.41}$$

Wigner functions $D^I_{MK}(\alpha)$ are defined to be matrix elements of the rotation operator (B.5):

$$\langle IM | A(\alpha) | IK\rangle = D^I_{MK}(\alpha) \qquad D^{I*}_{MK}(\alpha) = \langle IK | A^{-1}(\alpha) | IM\rangle \tag{B.42}$$

where α stands for the three Euler angles (α, β, γ). The Wigner functions are normalized to [225]:

$$D^{I_1}_{M_1K_1}(\alpha)\, D^{I_2}_{M_2K_2}(\alpha) = \sum_I \begin{bmatrix} I_1 & I_2 & I \\ M_1 & M_2 & M \end{bmatrix} \begin{bmatrix} I_1 & I_2 & I \\ K_1 & K_2 & K \end{bmatrix} D^I_{MK}(\alpha)$$

$$\int d\alpha D^{I*}_{MK}(\alpha)\, D^{I'}_{M'K'}(\alpha) = \frac{8\pi^2}{2I+1}\delta_{II'}\delta_{MM'}\delta_{KK'} \tag{B.43}$$

where $d\alpha$ is the invariant group measure, defined as in [225]:

$$\int d\alpha = \int_0^{2\pi} d\alpha \int_0^{\pi} d\beta \int_0^{2\pi} d\gamma\, |\det C| = \int_0^{2\pi} d\alpha \int_0^{\pi} \sin\theta d\theta \int_0^{2\pi} d\gamma \tag{B.44}$$

where $\begin{bmatrix} I & I_2 & I_1 \\ M & M_2 & M_1 \end{bmatrix} = \langle IM, I_2M_2 \mid I_1M_1\rangle$ are $SU(2)$ Clebsch–Gordan coefficients. They are listed in [7]. (In [216] the definitions of D^{I*}_{MK} and D^I_{MK} are interchanged.)

From (B.31) we see that:

$$\mathcal{R}_a D^{I*}_{MK}(\alpha) = \langle IK\, |\mathcal{R}_a A^{-1}(\alpha)|\, IM\rangle = \langle IK\, |A^{-1}(\alpha)\, j_a|\, IM\rangle \tag{B.45}$$

This relation shows that the differential operator $\mathcal{R}_a$ acts on the Wigner function $D^{I*}_{MK}(\alpha)$ in exactly the same way as the angular momentum operator j_a acts on the state $|IM\rangle$. Similarly, we show that:

$$\mathcal{L}_a D^{I*}_{MK}(\alpha) = \langle IK\, |\mathcal{L}_a A^{-1}(\alpha)|\, IM\rangle = \langle IK\, |j_a A^{-1}(\alpha)|\, IM\rangle \tag{B.46}$$

The differential operator $\mathcal{L}_a$ acts on the Wigner function $D^{I*}_{MK}(\alpha)$ in exactly the same way as the angular momentum operator j_a acts on the left on the state $\langle IK|$.

Let us define a normalized state $|IKM\rangle$ which, in the representation $\langle\alpha|$, is:

$$\langle \alpha \mid IKM\rangle = \sqrt{\frac{2I+1}{8\pi^2}}\langle IK\, |A^{-1}(\alpha)|\, IM\rangle = \sqrt{\frac{2I+1}{8\pi^2}} D^{I*}_{MK}(\alpha)$$

$$\langle IKM \mid I'K'M'\rangle = \int d\alpha \langle IKM \mid \alpha\rangle \langle \alpha \mid I'K'M'\rangle = \delta_{II'}\delta_{MM'}\delta_{KK'} \tag{B.47}$$

The relations (B.45) and (B.46), together with the relations (B.31), lead to the expressions:

$$\begin{gathered} \mathcal{R}_3 |IKM\rangle = M\, |IKM\rangle \qquad \mathcal{L}_3 |IKM\rangle = K\, |IKM\rangle \\ (\mathcal{R}_1 \pm i\mathcal{R}_2)\, |IKM\rangle = \sqrt{(I \mp M)(I \pm M + 1)}\, |IKM \pm 1\rangle \\ (\mathcal{L}_1 \pm i\mathcal{L}_2)\, |IKM\rangle = \sqrt{(I \pm K)(I \mp K + 1)}\, |IK \mp 1M\rangle \end{gathered} \tag{B.48}$$

The hamiltonian (B.13) can be diagonalized on the basis $|IKM\rangle$. The relations (B.48) yield the matrix elements of the first term, which involves only the intrinsic angular momentum operators $\mathcal{L}_a$. In the absence of the potential term $V(\alpha)$, the hamiltonian (B.13) would commute with the lab angular momentum operators $\mathcal{R}_a$ so that the eigenstates would bear the quantum numbers I and M. It is convenient to express the potential $V(\alpha)$ in terms of the Wigner functions $D^{I*}_{MK}(\alpha)$:

$$V(\alpha) = \sum_{IMK} c_{IKM} D^{I*}_{MK}(\alpha) \qquad C_{IKM} = \frac{2I+1}{8\pi^2} \int d\alpha D^{I}_{MK}(\alpha) V(\alpha) \tag{B.49}$$

Once the coefficients c_{IKM} are calculated, the matrix elements of the potential V are given by:

$$\langle I_1 K_1 M_1 |V| I_2 K_2 M_2 \rangle = \int d\alpha \langle I_1 K_1 M_1 \mid \alpha \rangle V(\alpha) \langle \alpha \mid I_2 K_2 M_2 \rangle$$

$$= \frac{8\pi^2}{\sqrt{(2I_1+1)(2I_2+1)}} \sum_{IKM} c_{IKM} \begin{bmatrix} I & I_2 & I_1 \\ M & M_2 & M_1 \end{bmatrix} \begin{bmatrix} I & I_2 & I_1 \\ K & K_2 & K_1 \end{bmatrix} \tag{B.50}$$

The coefficients appearing in (B.47), (B.49) and (B.50) depend on the normalization (B.44) of $d\alpha$.

Because the diagonalization on the basis $|IKM\rangle$ makes good use of the $SU(2)$ group properties, it is a far more efficient way to find the eigenstates and eigenvalues of the hamiltonian (B.13) than to diagonalize it numerically on the basis $|\alpha\rangle$.

The three Euler angles (α, β, γ) can be interpreted as coordinates which measure the orientation, relative to the lab frame, of a triad $(\boldsymbol{e}^1, \boldsymbol{e}^2, \boldsymbol{e}^3)$ of unit vectors which are fixed to a rotating body. Rotations of the body, relative to the lab, are generated by the lab frame angular momentum operators $\mathcal{R}_a$. The operators $\mathcal{L}_a$ can be thought of as the projections $\boldsymbol{e}^a.\boldsymbol{R}$ of the angular momentum operator on the body fixed frame. That is why they are called intrinsic angular momentum operators. The intrinsic angular momentum operators $\mathcal{L}_a = \boldsymbol{e}^a.\boldsymbol{R}$ are scalars with respect to rotations measured in the lab frame. They commute with the angular momentum operators $\mathcal{R}_a$.

B.5 The $SU(3)$ rotor

A useful introduction to the use of the $SU(3)$ group in particle physics is offered by [226]. Representations, Wigner functions and Clebsch–Gordan coefficients are listed in [227, 228]. Intrinsic and lab frame operators are given in [219] in terms of $SU(3)$ Euler angles. We limit the discussion to the tools required to diagonalize the $SU(3)$ rotor hamiltonian.

In the case of $SU(3)$, there are 8 generators j_a which obey the commutation rules (B.1), where the f_{abc} are the $SU(3)$ structure functions (A.1). In the fundamental representation, they can be written in the form $j_a = \frac{1}{2}\tau_a$, where the τ_a are the eight 3×3 Gell-Mann matrices listed in (A.22).

The first three operators j_1, j_2 and j_3 obey $SU(2)$ angular momentum commutation rules. They can be called isospin operators. The three isospin operators commute with the operator j_8, which is used to define a hypercharge operator Y. Thus we define:

$$\boldsymbol{T} = (T_1, T_2, T_3) = (j_1, j_2, j_3) \qquad Y = \frac{2}{\sqrt{3}} j_8 \tag{B.51}$$

It is useful to work with the following raising and lowering operators:

$$T_{\pm} = j_1 \pm j_2 \qquad U_{\pm} = j_6 \pm i j_7 \qquad V_{\pm} = j_4 \pm j_5 \tag{B.52}$$

A representation of the $SU(3)$ group is provided by the states $|(p,q)\,T, T_3, Y\rangle$ which are eigenstates of five commuting operators, namely the two Casimir operators C_2 and C_3 defined in (B.20), the total isospin $\boldsymbol{T}^2 = T_1^2 + T_2^2 + T_3^2$, its projection T_3 and the hypercharge Y. The eigenvalues of the two Casimir operators can be expressed in terms of two-positive integers p' and q':

$$C_2 = \frac{2}{3}\left(p'^2 + p'q' + q'^2\right) - 2$$

$$C_3 = \frac{1}{9}\left(p' + 2q'\right)\left(q' + 2p'\right)\left(p' - q'\right) + \left(p'^2 + p'q' + q'^2\right) - 3 \tag{B.53}$$

A multiplet (p', q') is composed of N states, where:

$$N = \frac{1}{2} p'q'\left(p' + q'\right) \tag{B.54}$$

It has become customary to denote the $SU(3)$ multiplets by the two non-negative integers $p = p' - 1$ and $q = q' - 1$. A given $SU(3)$ multiplet is denoted (p, q). The conjugate multiplet is obtained by exchanging p and q. Self-conjugate representations, such as the octet $(1,1)$, have $p = q$. Within any multiplet, that is, for given values of p and q, the representation is labeled by the eigenvalues of $\boldsymbol{T}^2$, T_3 and Y:

$$j_3 \,|(p,q)\,T, T_3, Y\rangle = T_3 \,|(p,q)\,T, T_3, Y\rangle$$

$$\frac{2}{\sqrt{3}} j_8 \,|(p,q)\,T, T_3, Y\rangle = Y \,|(p,q)\,T, T_3, Y\rangle$$

$$T_{\pm} \,|(p,q)\,T, T_3, Y\rangle = \sqrt{\left(T \mp T_3\right)\left(T \pm T_3 + 1\right)}\,|(p,q)\,T, T_3 \pm 1, Y\rangle \tag{B.55}$$

In the following, we assume that the states are normalized:

$$\langle (p,q)\,T, T_3, Y \mid (p', q')\,T', T_3', Y'\rangle = \delta_{pp'}\delta_{qq''}\delta_{TT'}\delta_{T_3T_3'}\delta_{YY'} \tag{B.56}$$

Multiplets are often denoted by their multiplicity N. The convention is to denote them by N when $p \geq q$ and to denote the conjugate multiplet (q, p) by $\bar{N}$. For

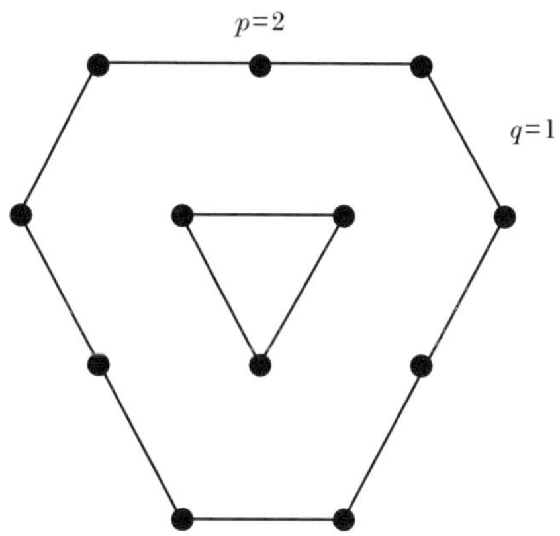

FIG. B.1. The $\{p, q\} = \{2, 1\}$ representation of the $SU(3)$ multiplet $N = 15$.

example, the three quarks u, d and s belong to the fundamental triplet representation $(1, 0) \equiv 3$. The antiquarks belong to the conjugate representation $(0, 1) \equiv \bar{3}$. The $N\,(939)\,,\Lambda\,(1116)\,,\Sigma\,(1193)$ and $\Xi\,(1318)$ are spin 1/2 baryons belonging to an octet $(1, 1) \equiv 8 \equiv \bar{8}$. The $\Delta\,(1232)\,,\Sigma\,(1385)\,,\Xi\,(1530)$ and $\Omega^-\,(1672)$ are spin 3/2 members of the decuplet $(3, 0) \equiv 10$. The vacuum is the singlet $(0, 0)$. The multiplicity is a convenient if not always faithful label of a multiplet.

A given representation $\{p, q\}$ is conveniently represented in the (Y, T_3) plane. The states $|(p, q)\,, T, T_3, Y\rangle$ are represented by points within and on the boundary of a convex polygon, which is symmetric with respect to rotations of 120° and which is symmetric with respect to the Y axis (Figure B.1) shows how the numbers p and q can be read off the polygon.

When p or q is zero, the polygon becomes a triangle. For such multiplets, each site contains one and only one state. There is only one state at each site on the boundary. In addition, the matrix elements of the shift operators $U_\pm$ and $V_\pm$ are the same as those of angular momentum operators, except that they act in different directions.

When both p and q are different from zero, the situation is complicated by the fact that the sites may contain more than one state. When several states are situated at the same site, the matrix elements of the shift operators are no longer given by simple analytic formulas. For not too large multiplets, they must be constructed by hand so to speak. The method is explained in [226]. The phases of the states are chosen such that [227]:

$$\langle T, T_3, Y\,|T_1 \pm iT_2|\,T', T_3', Y\rangle \geq 0 \qquad \langle T, T_3, Y\,|T_4 \pm iT_5|\,T', T_3', Y'\rangle \geq 0 \tag{B.57}$$

For the octet $(1, 1)$, which has two states in its center point (one isosinglet and one member of the isotriplet), the matrix elements of the shift operators are

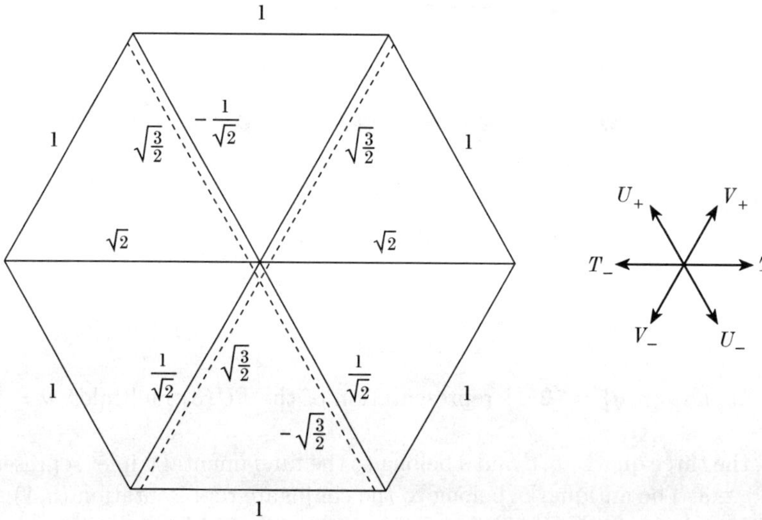

FIG. B.2. Matrix elements of the $SU(3)$ shift operators (B.52). The dashed lines are the matrix elements linking the isosinglet state to the isodoublets.

shown in Figure B.2.

Wigner functions for the $SU(3)$ group are defined in the same way as in $SU(2)$, namely:

$$\langle N\nu | A(\alpha) | N'\nu' \rangle = \delta_{NN'} D^N_{\nu\nu'}(\alpha) \qquad D^{N*}_{\nu\nu'}(\alpha) = \langle N'\nu' | A^{-1}(\alpha) | N\nu \rangle \tag{B.58}$$

where we use the abridged notation:

$$N \equiv (p, q) \qquad \nu \equiv (T, T_3, Y) \tag{B.59}$$

They satisfy the relations:

$$D^{N_1}_{\nu_1\nu_1'}(\alpha)\, D^{N_2}_{\nu_2\nu_2'}(\alpha) == \sum_{N\nu} \begin{bmatrix} N_1 & N_2 & N \\ \nu_1 & \nu_2 & \nu \end{bmatrix} \begin{bmatrix} N_1 & N_2 & N \\ \nu_1' & \nu_2' & \nu' \end{bmatrix} D^N_{\nu\nu'}(\alpha)$$

$$\int d\alpha D^{N_1}_{\nu_1\nu_1'}(\alpha)\, D^{N_2*}_{\nu_2\nu_2'}(\alpha) = \frac{1}{N} \delta_{N,N'} \delta_{\nu_1,\nu_2} \delta_{\nu_1',\nu_2'} \tag{B.60}$$

where $\begin{bmatrix} N_1 & N_2 & N \\ \nu_1 & \nu_2 & \nu \end{bmatrix}$ are $SU(3)$ Clebsch–Gordan coefficients. On the right hand side, N stands for the multiplicity (B.54)of the representation. The properties of the $SU(3)$ Wigner functions and the Clebsch–Gordan coefficients are listed in [227]. The normalization (B.60) depends on the definition of the group invariant measure $d\alpha$. The result (B.60)conforms to the conventions of [227]. For $SU(2)$, the definition of $d\alpha$ used in [227] is a factor $8\pi^2$ smaller than the corresponding definition in [216, 225]. We can define an orthonormal basis of states:

$$\langle \alpha \mid N\nu\nu' \rangle = \sqrt{N} \left\langle N\nu \left| A^{-1}(\alpha) \right| N\nu' \right\rangle = \sqrt{N} D^{N*}_{\nu'\nu}(\alpha) \qquad \text{(B.61)}$$

The relations (B.45) and (B.46) apply also to the $SU(3)$ case. One can thus deduce how the intrinsic and lab frame operators $\mathcal{L}_a$ and $\mathcal{R}_a$ act on the basis states (B.61), from the way the $SU(3)$ operators act on the states $\langle N\nu|$ and $|N\nu'\rangle$ respectively. The rotor hamiltonian (B.13) can therefore be diagonalized on the $SU(3)$ basis (B.61). The generators j_a form an octet, so that:

$$A^{-1}(\alpha)\, j_a A(\alpha) = D^{(8)*}_{ab}(\alpha)\, j_b \qquad \text{(B.62)}$$

APPENDIX C

BOSONIZATION

C.1 Gaussian transformations and Wick theorems

Let us calculate:

$$e^{\frac{1}{2}G^2F^2} = \int_{-\infty}^{\infty} d\lambda \delta\left(\lambda - iF\right) e^{-\frac{G^2}{2}\lambda^2} = \int_{-\infty}^{\infty} dS \frac{1}{2\pi} \int_{-\infty}^{\infty} d\lambda e^{-iS(\lambda - iF)} e^{-\frac{G^2}{2}\lambda^2} \tag{C.1}$$

We integrate first over λ to get:

$$e^{\frac{1}{2}G^2F^2} = \frac{1}{\sqrt{2\pi}G} \int_{-\infty}^{\infty} dS e^{-\frac{S^2}{2G^2} - SF} = \frac{\int_{-\infty}^{\infty} dS e^{-\frac{1}{2G^2}S^2 - SF}}{\int_{-\infty}^{\infty} dS e^{-\frac{1}{2G^2}S^2}} \tag{C.2}$$

In the expressions above, F is a given operator. Equation (C.2) expresses a quadratic form of the operator F in terms of a linear form, at the expense of introducing the extra variable S. If we expand both sides of the expression (C.2) in powers of F, we generate a Wick theorem for gaussian integrals:

$$\left\langle S^{2n+1} \right\rangle \equiv \frac{\int_{-\infty}^{\infty} dS e^{-\frac{1}{2G^2}S^2} S^{2n+1}}{\int_{-\infty}^{\infty} dS e^{-\frac{1}{2G^2}S^2}} = 0 \qquad \left\langle S^2 \right\rangle \equiv \frac{\int_{-\infty}^{\infty} dS e^{-\frac{1}{2G^2}S^2} S^2}{\int_{-\infty}^{\infty} dS e^{-\frac{1}{2G^2}S^2}} = G^2$$

$$\left\langle S^4 \right\rangle \equiv \frac{\int_{-\infty}^{\infty} dS e^{-\frac{1}{2G^2}S^2} S^4}{\int_{-\infty}^{\infty} dS e^{-\frac{1}{2G^2}S^2}} = 3\left\langle S^2 \right\rangle^2 = 3G^4, \ ... \tag{C.3}$$

Proceeding in the same way, we can also calculate:

$$e^{-\frac{1}{2}G^2F^2} = \int_{-\infty}^{\infty} d\lambda \delta\left(F - \lambda\right) e^{-\frac{1}{2}G^2\lambda^2} = \frac{\int_{-\infty}^{\infty} dS e^{-\frac{1}{2G^2}S^2 - iSF}}{\int_{-\infty}^{\infty} dS e^{-\frac{1}{2G^2}S^2}} \tag{C.4}$$

The calculations above can be applied to more general functions of F. However, only a gaussian function of F leads to a gaussian integral over λ. In other cases, the integral over λ can be performed by the stationary phase method.

The identities (C.2) and (C.4) can be extended to functional integrals. We define a real integration field $S(x) = \langle x \mid S\rangle = \langle S \mid x\rangle$ and an interaction matrix $\langle x|V|y\rangle = \langle y|V|x\rangle$. The identity:

$$\int D(S)\, e^{-\frac{1}{2}(S+VF)V^{-1}(S+VF)} = \int D(S)\, e^{-\frac{1}{2}SV^{-1}S} \tag{C.5}$$

leads to the identity:

$$e^{\frac{1}{2}FVF} = \frac{\int D(S)\, e^{-\frac{1}{2}SV^{-1}S} e^{-SF}}{\int D(S)\, e^{-\frac{1}{2}SV^{-1}S}} \tag{C.6}$$

Expansion of both sides in powers of F leads to a Wick theorem for which the only non-zero contraction is equal to:

$$\langle S(x)\, S(y)\rangle \equiv \frac{\int D(S)\, e^{-\frac{1}{2}SV^{-1}S} S(x)\, S(y)}{\int D(S)\, e^{-\frac{1}{2}SV^{-1}S}} = \langle x\, |V|\, y\rangle \tag{C.7}$$

The identity (C.6) and the Wick theorem provide the basis for bosonization as well as for the expansion of the partition function in terms of Feynman graphs.

REFERENCES

1. S. Godfrey and N. Isgur. *Phys. Rev. D32*, page 189, 1985.
2. S. Capstick and N. Isgur. *Phys. Rev. D34*, page 2809, 1986.
3. F. Schöberl, W. Lucha, and D. Gromes. *Phys. Rep. 200*, page 127, 1991.
4. L. Glozman and D. Riska. *Phys. Rep. 268*, page 263, 1996.
5. C. Itzykson and J. B. Zuber. *Quantum Field Theory.* McGraw Hill, 1980.
6. M. A. Shifman, A. I. Vainstein, M. B. Voloshin, V. A. Novikov, L. B. Okun, and V. I. Zakharov. *Phys. Rep. 41*, page 1, 1978.
7. Particle Data 1994. *Phys. Rev. D50*, page 1175, 1994.
8. Particle Data 1996. *Phys. Rev. D54*, page 1, 1996.
9. E. V. Shuryak. *Rev. Mod. Phys. 65*, page 1, 1993.
10. S. Huang, M.-C. Chu, J. M. Grandy, and J. W. Negele. *Phys. Rev. D48*, page 3340, 1993.
11. A. A. Grigorian and A. E. Tyugu. *Phys. Lett. B216*, page 198, 1989.
12. AA. Grigorian and E. T. Tyugu. *Phys. Lett. B222*, page 538 (Erratum), 1989.
13. J. Praschifka, R. T. Cahill, and C. J. Burden. *Austr. Jour. Phys. 42*, pages 161,171, 1989.
14. R. D. Ball. *Int. Journ. Mod. Phys. A5*, page 4391, 1990.
15. U. Vogl and W. Weise. *Prog. Part. Nucl. Phys. 27*, page 195, 1991.
16. H. Reinhardt, D. Ebert, and M. K. Volkov. *Prog. Part. Nucl. Phys. 33*, page 1, 1994.
17. T. Goldman and R. W. Haymaker. *Phys. Rev. D24*, page 724, 1981.
18. R. Ball. In M. Jezabek and M. Praszalowicz, editors, *Workshop on Skyrmions and Anomalies (Cracow 1987).* World Scientific, 1987.
19. R. T. Cahill, C. D. Roberts, and J. Praschifka. *Ann. Phys. (NY) 188*, page 20, 1988.
20. R. T. Cahill, J. Prashifka, and C. D. Roberts. *Int. Jour. Mod. Phys. A4*, page 4929, 1989.
21. C. D. Roberts. In *QCD Vacuum Structure*, page 114. World Scientific, 1992.
22. C. D. Roberts and A. G. Williams. *Prog. Part. and Nucl. Phys. 33*, page 475, 1994.
23. R. Alkofer and H. Reinhardt. *Zeit. Phys. A343*, page 79, 1992.
24. H. Kleinert. Hadronization of quark theories. In A. Zichichi, editor, *Understanding the Fundamental Constituents of Matter.* Plenum Press, 1976.
25. E. V. Shuryak. *Phys. Rep. 115*, page 152, 1984.
26. D. I. Diakonov and V. Y. Petrov. *Nucl. Phys. B272*, page 457, 1986.
27. D. Diakonov. Chiral symmetry breaking by instantons. Lectures given at the Enrico Fermi School in Physics, Varenna, June 27 - July 7, 1995.
28. L. Y. Glozman. Baryon structure and the chiral symmetry of QCD. In H. Latal

and W. Schweiger, editors, *Perturbative and Non Perturbative Aspects of Quantum Field Theory.* Springer, 1996.

29. U. Vogl, S. Klimt, M. Lutz, and W. Weise. *Nucl. Phys. A516*, page 429 and 469, 1990.
30. G. Ripka, M. Jaminon, R. Mendez-Galain, and P. Stassart. *Nucl. Phys. A537*, page 418, 1992.
31. E. Shuryak. *Nucl. Phys. B203*, pages 93,116,140, 1982.
32. G. 't Hooft. *Phys. Rep. 142*, page 357, 1986.
33. J. J. M. Verbaarschot, R. Alkofer, M. A. Nowak, and I. Zahed. *Phys. Lett. B233*, page 205, 1989.
34. S. Huang, M.-C. Chu, J. M. Grandy, and J. W. Negele. *Phys. Rev. D49*, page 6039, 1994.
35. D. Blaschke, S. Schmidt, and Y. Kalinovsky. *Phys. Rev. C50*, page 435, 1994.
36. L. Münchow, V. N. Pervushin, G. Röpke, D. Blaschke, Y. Kalinovsky, and S. Schmidt. *Nucl. Phys. A586*, page 711, 1995.
37. M. Buballa and S. Krewald. *Phys. Lett. B294*, page 19, 1992.
38. R. D. Bowler and M. C. Birse. *Nucl. Phys. A582*, page 655, 1995.
39. Y. Nambu and G. Jona-Lasinio. *Phys. Rev. 122*, page 345, 1961.
40. Y. Nambu and G. Jona-Lasinio. *Phys. Rev. 124*, page 246, 1961.
41. R. P. Feynman and A. R. Hibbs. *Quantum Mechanics and Path Integrals.* McGraw Hill, 1965.
42. J. P. Blaizot and G. Ripka. *Quantum Theory of Finite Systems.* MIT Press, 1986.
43. A. Vainstein, M. A. Shifman, and V. Zacharov. *Nucl. Phys. B147*, pages 385,448, 1979.
44. H. Rubinstein, L. I. Reinders, and S. Yazaki. *Phys. Rep. 87*, page 77, 1982.
45. S. Narison. *QCD Sum Rules.* World Scientific (Singapore), 1990.
46. R. Gupta and T. Bhattacharya. Light quark masses from lattice qcd. Los Alamos preprint LAUR-96-1840, hep-lat/9605039, 1996.
47. I. J. R. Aitchison. *An Informal Introduction to Gauge Field Theories.* Cambridge University Press, 1982.
48. R. Tegen, V. Dmitranisovic, H. J. Schulze, and R. H. Lemmer. *Phys. Rev. D52*, page 2855, 1995.
49. E. Witten. *Nucl. Phys. B156*, page 269, 1979.
50. G. Veneziano. *Nucl. Phys. B159*, page 213, 1979.
51. H. Reinhardt and R. Alkofer. *Phys. Lett. B207*, page 482, 1988.
52. T. Hatsuda and T. Kunihiro. *Phys. Rep. 247*, page 221, 1994.
53. J. Schechter. *Phys. Rev. D21*, page 3393, 1980.
54. A. Blotz. Strangeness in the semibosonized Nambu–Jona-Lasinio model. PhD thesis, Ruhr-Universität Bochum (Germany), 1994.
55. M. Jaminon and B. Van den Bossche. *Nucl. Phys. A582*, page 517, 1995.
56. M. B. Voloshin and M. A. Schifman. *Sov. Journ. Nucl. Phys. 45*, page 292, 1987.
57. N. Isgur and M. B. Wise. *Phys. Lett. B237*, page 527, 1990.

58. H. Georgi, A. F. Falk, and B. Grinstein. *Nucl. Phys. B343*, page 1, 1990.
59. R. Friedrich, D. Ebert, T. Feldman, and H. Reinhardt. *Nucl. Phys. B434*, page 619, 1995.
60. G. 't Hooft. *Nucl. Phys. B75*, page 461, 1974.
61. E. Witten. *Nucl. Phys. B160*, page 57, 1979.
62. K. Goeke, P. Pobylitsa, V. Petrov, M. Wakamatsu, C. Christov, A. Blotz, and T. Watabe. *Phys. Lett. B325*, page 467, 1994.
63. R. J. Oakes, M. Gell-Mann, and B. Renner. *Phys. Rev. 175*, page 2194, 1968.
64. G. Ripka, M. Jaminon, and P. Stassart. *Phys. Lett. B227*, page 191, 1989.
65. J. Stern, M. Knecht, B. Moussallam, and N. H. Fuchs. *Nucl. Phys. B457*, page 513, 1995.
66. L. Montanet. *Nucl. Phys. 39B,C (Proc. Suppl.)*, page 281, 1995.
67. L. Burakowsky and L. P. Horwitz. *Nucl. Phys. A609*, page 585, 1996.
68. R. Ball. *Phys. Rep. 182*, page 1, 1989.
69. C. D. Roberts, J. Praschifka, and R. T. Cahill. *Phys. Rev. D36*, page 209, 1987.
70. J. Terning, B. Holdom, and K. Verbeek. *Phys. Lett. B232*, page 351, 1989.
71. R. D. Ball and G. Ripka. In C. Souza, C. Fiolhais, M. Fiolhais, and J. N. Urbano, editors, *Many Body Physics (Coimbra 1993)*. World Scientific, 1993.
72. P. Tandy, F. Frank, and G. Fai. *Phys. Rev. C43*, page 2808, 1991.
73. F. Gross and J. Milana. *Phys. Rev. D45*, page 969, 1992.
74. C. Roberts, C. Burden, and A. Williams. *Phys. Lett. B285*, page 347, 1992.
75. H. Reusch, M. Stingl, U. Häbel, R. Könning, and S. Wizard. *Zeit. Phys. A336*, pages 423,435, 1990.
76. M. Frank and C. Roberts. *Phys. Rev. C53*, page 390, 1996.
77. K. Langfeld and M. Rho. *Nucl. Phys. A596*, page 451, 1996. Saclay preprint T95/070.
78. K. Langfeld. *Nucl. Phys. A602*, page 375, 1996.
79. T. Feldmann, D. Ebert, and H. Reinhardt. *Phys. Lett. B388*, page 154, 1996.
80. G. V. Efimov and O. A. Mogilevsky. *Nucl. Phys. B44*, page 541, 1972.
81. Y. Kalinovsky and C. Weiss. *Zeit. Phys. C63*, page 275, 1994.
82. M. K. Volkov and C. Weiss. A chiral lagrangian for excited pions. University of Bochum preprint RUB-TPH-12/96, hep-ph/9608347.
83. Th. Meissner, A. Blotz, F. Döring, and K. Goeke. *Phys. Lett. B251*, page 235, 1990.
84. J. Schwinger. *Phys. Rev. 82*, page 664, 1951.
85. E. N. Nikolov, W. Broniowski, G. Ripka, and K. Goeke. *Zeit. Phys. A354*, page 421, 1996.
86. R. Tegen, V. Dmitrasinovic, H. Schulze, and R. Lemmer. *Annals Phys. 238*, page 332, 1995.
87. C. Christov, G. Ripka, E. Nikolov, W. Broniowski, and K. Goeke. *Nucl. Phys. A608*, page 411, 1996.
88. W. Pauli and F. Villars. *Rev. Mod. Phys. 21*, page 434, 1949.
89. E. Ruiz Arriola, C. Schüren, F. Döring, and K. Goeke. *Nucl. Phys. A565*, page 687, 1993.

90. E. Ruiz Arriola. *Phys. Lett. B253*, page 430, 1991.
91. E. Ruiz Arriola, C. Schüren, and K. Goeke. *Nucl. Phys. A547*, page 612, 1992.
92. E. V. Shuryak and J. J. M. Verbaarschot. *Nucl. Phys. B410*, page 37 and 55, 1993.
93. E. V. Shuryak, T. Schäfer, and J. J. M. Verbaarschot. *Nucl. Phys. B412*, page 143, 1994.
94. I. Montvay and G. Münster. *Quantum Fields on a Lattice.* Cambridge University Press, 1994.
95. G. Ripka and S. Kahana. *Phys. Rev. D36*, page 1233, 1987.
96. R. Perry. *Phys. Lett. B199*, page 489, 1987.
97. S. Kahana and G. Ripka. *Phys. Lett. B278*, page 11, 1992.
98. R. M. Davidson and E. Ruiz Arriola. *Phys. Lett. B359*, page 273, 1995.
99. C. Christov. Private communication.
00. S. Coleman and E. Weinberg. *Phys. Rev. D7*, page 1888, 1973.
01. G. Baym. *Phys. Rev. 127*, page 1391, 1962.
02. C. De Dominicis and P. C. Martin. *Jour. Math. Phys. 5*, page 14 and 31, 1964.
03. D. J. Thouless. *Quantum Mechanics of Many-Body Systems.* Academic Press, 1961.
04. G. E. Brown. *Unified Theory of Nuclear Models and Forces.* North Holland, 1967.
05. W. Florkowski and W. Broniowski. *Phys. Lett. B386*, page 62, 1996.
06. P. Zhuang, J. Hüfner, S. P. Klevansky, and H. Voos. *Ann. Phys. 234*, page 225, 1994.
07. J. Hüfner, P. Zhuang, and S. P. Klevansky. *Nucl. Phys. A546*, page 525, 1994.
08. K. Akama. *Phys. Rev. Lett. 76*, page 184, 1996.
09. C. D. Roberts, A. Bender, and L. V. Smekal. *Phys. Lett. B380*, page 7, 1996.
10. S. Adler. *Phys. Rev. 177*, page 2426, 1969.
11. W. A. Bardeen. *Phys. Rev. 184*, page 1848, 1969.
12. R. Jackiw. In R. Jackiw, S. B. Treiman, and D. F. Gross, editors, *Lectures in Current Algebra and its Applications*, page 97. Princeton University Press, 1972.
13. B. Hiller, A. H. Blin, and M. Schaden. *Zeit. Phys. A331*, page 75, 1988.
14. E. Ruiz Arriola and L. L. Salcedo. *Nucl. Phys. A590*, page 703, 1995.
15. A. N. Ivanov, D. Ebert, and M. K. Volkov. *Fortschr. Phys. 37*, page 513, 1989.
16. S. P. Klevansky. *Rev. Mod. Phys. 64*, page 649, 1992.
17. R. Alkofer and H. Reinhardt. *Chiral Quark Dynamics.* Springer Verlag (Berlin–Heidelberg), 1995.
18. J. Bijnens. *Phys. Rep. 265*, page 369, 1996.
19. S. Weinberg. *Physica A96*, page 327, 1979.
20. J. Gasser and H. Leutwyler. *Nucl. Phys. B250*, pages 465,517,539, 1985.
21. B. R. Holstein. Low energy tests of chiral perturbation theory. Talk given at the PANIC 96 meeting in Williamsburg, Virginia, hep-ph/9606228.
22. A. Manohar. Effective field theories. Lectures at the Schmalding Winter School, March 1996, University of California at San Diego preprint YCSD/PTH 96-

04, hep/ph9606222.
123. C. Bruno, J. Bijnens, and E. DeRafael. *Nucl. Phys. B390*, page 501, 1993.
124. E. Tudor, I. Aitchison, C. Frazer, and J. Zuk. *Phys. Lett. B165*, page 162, 1985.
125. L.-H. Chan. *Phys. Rev. Lett. 39*, page 1124, 1977.
126. H. Georgi. *Phys. Rev. Lett. 63*, page 1917, 1989.
127. G. E. Brown and M. Rho. *Phys. Lett. B338*, page 301, 1994.
128. B. D. Serot and J. D. Walecka. *Advances in Phys. 16*, 1986.
129. R. J. Furnstahl, T. D. Cohen, and D. K. Griegel. *Phys. Rev. C45*, page 883, 1992.
130. H. Leutwyler, J. Gasser, and M. E. Sanio. *Phys. Lett. B253*, page 1991, 1991.
131. R. J. Furnsdahl and C. J. Horowitz. *Nucl. Phys. A485*, page 579, 1988.
132. R. Mendez-Galain, M. Jaminon, and P. Stassart. *Nucl. Phys. A556*, page 509, 1993.
133. T. Matsui and B. D. Serot. *Ann. Phys. 144*, page 107, 1982.
134. A. Pantziris, L. S. Celenza, and C. M. Shakin. *Phys. Rev. C45*, page 205, 1992.
135. M. Jaminon and G. Ripka. *Nucl. Phys. A564*, page 505, 1993.
136. V. Bernard and U.-G. Meissner. *Nucl. Phys. A489*, page 647, 1988.
137. S. Klimt, M. Lutz, and W. Weise. *Nucl. Phys. A542*, page 521, 1992.
138. M. Jaminon and R. Mendez-Galain. *Nucl. Phys. A564*, page 542, 1993.
139. M. Jaminon and B. Van den Bossche. *Nucl. Phys. A582*, page 517, 1995.
140. J. Gasser and H. Leutwyler. *Phys. Lett. B184*, page 83, 1987.
141. L. Dolan and R. Jackiw. *Phys. Rev. D9*, page 3320, 1974.
142. H. E. Haber and H. A. Weldon. *J. Math. Phys. 23*, page 1852, 1982.
143. B. Van den Bossche. Extension of the Nambu Jona-Lasinio model to finite temperatures and densities. PhD thesis, Université de Liège, Faculté des Sciences, 1996.
144. M. Abramowitz and I. A. Stegun. *Handbook of Mathematical Functions*. Dover Publications, Inc. ,New-York, 1965.
145. M. Jaminon and B. Van Den Bossche. *Zeit. Phys. C64*, page 339, 1994.
146. R. L. Jaffe, V. Bernard, and U.-G. Meissner. *Nucl. Phys. B308*, page 753, 1988.
147. U. Vogl, S. Klimt, M. Lutz, and W. Weise. *Phys. Lett. B249*, page 386, 1990.
148. J. P. Blaizot and R. Mendez-Galain. *Phys. Lett. B271*, page 32, 1991.
149. M. Jaminon and B. Van den Bossche. *Nucl. Phys. A567*, page 865, 1994.
150. Y. L. Kalinowsky, P. Petrov, S. Schmidt, D. Blaschke, M. Jaminon, and B. Van den Bossche. *Nucl. Phys. A588*, page 517, 1995.
151. S. P. Klevansky, P. Rehberg, and J. Hüfner. *Phys. Rev. C53*, page 410, 1996.
152. V. Yu. Petrov, D. I. Diakonov, and M. Praszalowicz. *Nucl. Phys. B323*, page 53, 1989.
153. F. Grümmer, P. Sieber, Th. Meissner, and K. Goeke. *Nucl. Phys. A547*, page 459, 1992.
154. R. Wünsch, P. Sieber, F. Grümmer, Th. Meissner, G. Ripka, and K. Goeke. *Phys. Lett. B299*, page 183, 1993.
155. A. Manohar and H. Georgi. *Nucl. Phys. B234*, page 189, 1984.

56. M. Praszalowicz, P. Sieber, and K. Goeke. *Nucl. Phys. A569*, page 629, 1994.
57. A. Duncan, J. Collins, and S. Joglekar. *Phys. Rev. D16*, page 438, 1977.
58. E. Golowich, J. F. Donoghue, and B. R. Holstein. *Dynamics of the Standard Model.* Cambridge Monographs on Particle Physics, Nuclear Physics and Cosmology, Cambridge University Press, 1992.
59. M. Jaminon and G. Ripka. *Nucl. Phys. A564*, page 505, 1993.
60. R. Johnson, H. Gomm, P. Jain, and J. Schechter. *Phys. Rev. D33*, page 801, 1986.
61. G. Ripka and M. Jaminon. *Ann. Phys. 218*, page 51, 1992.
62. K. Kusaka and W. Weise. *Zeit. Phys. A343*, page 229, 1992.
63. B. Van den Bossche, J. Cugnon, and M. Jaminon. *Nucl. Phys. A598*, page 515, 1996.
64. J. Ellis, B. A. Campbell, and K. A. Olive. *Nucl. Phys. B345*, page 57, 1990.
65. G. E. Brown and M. Rho. *Phys. Rev. Lett. 66*, page 2720, 1991.
66. R. Johnson, R. Jain, and J. Schechter. *Phys. Rev. D38*, page 1571, 1988.
67. K. Kusaka and W. Weise. *Nucl. Phys. A580*, page 383, 1994.
68. M. Gell-Mann and M. Levy. *Nuov. Cim. 16*, page 705, 1960.
69. G. Ripka, S. Kahana, and V. Soni. *Nucl. Phys. A415*, page 351, 1984.
70. M. Birse and M. K. Banerjee. *Phys. Lett. B136*, page 284, 1984.
71. W. Broniowski and M. Banerjee. *Phys. Rev. D34*, page 849, 1986.
72. M. C. Birse. *Prog. Part. Nucl. Phys. 25*, page 1, 1990.
73. J. A. McGovern and M. C. Birse. *Nucl. Phys. A506*, pages 367,392, 1990.
74. M. A. Nowak P. O. Mazur and M. Praszalowicz. *Phys. Lett. B147*, page 137, 1984.
75. F. Beck, J. Hartmann, and W. Bentz. *Phys. Rev. C50*, page 3088, 1994.
76. O. Nachtmann, G. Chanfray, and H. J. Pirner. *Phys. Lett. 147B*, page 249, 1984.
77. A. Schuh and H. J. Pirner. *Phys. Lett. B173*, page 19, 1986.
78. A. W. Thomas. *Adv. Nucl. Phys. 13*, page 1, 1983.
79. I. Duck. *Phys. Rev. D34*, page 1493, 1986.
80. M. K. Banerjee. *Prog. Part. Nucl. Phys. 31*, page 77, 1993.
81. J. A. McGovern. *Nucl. Phys. A533*, page 553, 1991.
82. M. Bae and J. A. McGovern. *Jour. Phys. G22*, page 199, 1996.
83. T. D. Lee. *Particle Physics and Introduction to Field Theory.* Harwood Academic Press (New York), 1981.
84. L. Wilets. *Nontopological Solitons.* World Scientific (Singapore), 1989.
85. H. Kim, P. Pobilitsa, T. Watabe, Th. Meissner, E. Ruiz Arriola, C. Christov, A. Blotz, and K. Goeke. *Prog. Part. Nucl. Phys. 37*, page 1, 1996.
86. H. Reinhardt, R. Alkofer, and H. Weigel. *Phys. Rep. 265*, page 139, 1996.
87. K. Johnson, C. B. Thorn, A. Chodos, R. L. Jaffe, and V. F. Weisskopf. *Phys. Rev. D9*, page 3471, 1974.
88. K. Johnson, A. Chodos, R. L. Jaffe, and C. B. Thorn. *Phys. Rev. D10*, page 2599, 1974.
89. S. Kahana and G. Ripka. *Nucl. Phys. A429*, page 462, 1984.

190. P. Sieber, J. Schaldach, D. Diakonov, M. Polyakov, and K. Goeke. *Phys. Rev. D53*, page 3366, 1996.
191. D. Diakonov, J. Schaldach, P. Sieber, and K. Goeke. *Phys. Rev. D54*, page 2814, 1996.
192. B. Moussallam. *Phys. Rev. D40*, page 3430, 1989.
193. M. Li and R. J. Perry. *Phys. Rev. D37*, page 1670, 1988.
194. E. Witten. *Phys. Lett. B117*, page 324, 1982.
195. V. Patodi, M. F. Atiyah, and I. Singer. *Math. Proc. Camb. Philos. Soc. 79*, page 71, 1976.
196. M. F. Atiyah and I. M. Singer. *Annals of Math. 93*, page 119, 1971.
197. T. H. R. Skyrme. *Nucl. Phys. 31*, page 556, 1962.
198. E. Witten. *Nucl. Phys. B223*, pages 422,433, 1983.
199. J. Goldstone and F. Wilczek. *Phys. Rev. Lett. 47*, page 986, 1981.
200. I. J. R. Aitchison and C. M. Fraser. *Phys. Rev. D31*, page 2605, 1985.
201. I. Zahed and G. Brown. *Phys. Rep. 142*, page 1, 1986.
202. G. Holzwarth, editor. *Baryons as Skyrme Solitons.* World Scientific (Singapore), 1993.
203. C. R. Nappi, G. S. Adkins, and E. Witten. *Nucl. Phys. B228*, page 552, 1983.
204. M. Gaudin, 1984. Private communication.
205. K. Goeke, N. W. Park, V. Petrov, A. Blotz, D. Diakonov, and P. V. Pobylitsa. *Nucl. Phys. A555*, page 765, 1993.
206. R. Alkofer, H. Weigel, and H. Reinhardt. *Nucl. Phys. A582*, page 484, 1995.
207. I. J. R. Aitchison and G. Ripka. *Nucl. Phys. A606*, 1996.
208. E. Ruiz Arriola, T. Watabe, F. Döring, C. Schüren, and K. Goeke. *Nucl. Phys. A603*, page 415, 1996.
209. R. Alkofer, H. Weigel, U. Zückert, and H. Reinhardt. *Nucl. Phys. A585*, page 513, 1995.
210. J. P. Blaizot and G. Ripka. *Phys. Rev. D38*, page 1556, 1988.
211. E. Guadagnini. *Nucl. Phys. B236*, page 35, 1984.
212. R. Alkofer, H. Weigel, and H. Reinhardt. *Nucl. Phys. B387*, page 638, 1992.
213. C. Callan and I. Klebanov. *Nucl. Phys. B262*, page 365, 1988.
214. R. Alkofer, H. Weigel, and H. Reinhardt. *Nucl. Phys. A576*, page 477, 1994.
215. R. Rajaraman. *Solitons and Instantons.* North Holland, 1982.
216. A. Bohr and B. R. Mottelson. *Nuclear Structure.* W. A. Benjamin, Inc. , 1969.
217. J. Schechter and H. Weigel. *Phys. Rev. D51*, page 6296, 1995.
218. K. Goeke, C. V. Christov, and P. V. Pobylitsa. *Phys. Rev. C52*, page 425, 1995.
219. H. Yabu and K. Ando. *Nucl. Phys. B301*, page 601, 1988.
220. A. Sudbery, A. J. Macfarlane, and P. H. Weisz. *Comm. Math. Phys. 11*, page 77, 1968.
221. A. J. Macfarlane. *Comm. Math. Phys. 11*, page 91, 1968.
222. M. Gell-Mann. *Phys. Rev. 125*, page 1067, 1962.
223. L. L. Okun. *Leptons and Quarks.* North Holland, 1982.
224. T.-P. Cheng and L.-F. Li. *Gauge Theory of Elementary Particles.* Clarendon

Press (Oxford), 1984.
225. A. Messiah. *Quantum Mechanics.* North Holland, 1961.
226. S. Gasiorowicz. *Elementary Particle Physics.* Wiley, 1966.
227. J. J. De Swart. *Rev. Mod. Phys. 35*, page 916, 1963.
228. S. J. Chilton, P. McNamee, and F. F. Chilton. *Rev. Mod. Phys. 36*, page 1005, 1964.

INDEX